SubjectMath.com Practice Test #2

A Full Practice Test For the Subject Math Exam
www.SubjectMath.com

First Edition

Intoroduction

Testimonial

By Prof. Amir Alexander (UCLA):

Every year thousands of aspiring mathematicians from the U.S. and beyond flood American universities with applications for graduate studies. Almost all of them are required to take the GRE subject exam in mathematics, and the results are critical to their success. How well they do on this test could determine whether they will be admitted to the program of their choice, or even accepted at all. Yet despite the high stakes, it is nearly impossible to find study materials for the test: there are no preparation courses, and only a single sample test is provided by the exam's administrators. How is one to study for this test, which could shape one's career for decades to come? Gilad Pagi and the GP group have the answer. Their website at SubjectMath.com offers a systematic study course for the Mathematics subject exam, the only one of its kind. In dozens of clear online lectures, filled with examples and study problems, the course covers all the topics included in the exam - from the calculus to analysis to algebra and beyond. A series of practice test booklets that precisely reproduce the content, format, and conditions of the actual test accompany the course. A student that has passed through this study course will go into the exam room confident and fully-prepared, and immensely improve his or her chances of success. If you are a student planning to take the GRE subject exam in mathematics, take notice: This is one course you cannot afford to miss.

Prof. Amir Alexander, UCLA

Preface

This book is a full practice test simulating the GRE® (Graduate Record Examinations) Subject Exam in Mathematics administered by ETS (Educational Testing Service). This practice test belongs to a series of tests that are part of our online preparation course (see www.SubjectMath.com). Each test is published individually. It is highly recommended that these tests be completed in correspondence with the course as solutions may reference results and examples further discussed within the course lectures.

If you are applying to graduate math programs in the US, you must excel in this test in order to be accepted. Alas, relevant materials are scarce and you have no one to tutor you throughout the exam. That's where we, the team in GP Group, come in.

This practice test was written as a part of our preparation course, encompassing all aspects of the subject exam. The team at GP Group composed a series of tests similar to the actual math subject exam in many respects:

- ✓ The test consists of 66 multiple choice questions.

- ✓ The content of the questions is taken from the official syllabus of the test.

- ✓ The style of the questions is similar to the questions in the official example test, published by ETS

- ✓ The distribution of the topics among the questions corresponds to the distribution as published by ETS and as seen in the published example test.

- ✓ The printing layout, including the space for scratch work, matches the real exam (as published).

- ✓ The test was designed to be taken in the same time frame and conditions as the real exam.

However, all the questions are original and are not published anywhere other than with the official course materials. Considering the scarcity of the prep materials for the subject exam, this book will improve your potential score significantly and, together with the online course, provide a well rounded preparation for the test.

At the date of publication, this series of books are the only practice exams not published by ETS which possess all the above features. Considering the scarcity of the prep materials for the subject exam, this book will improve your potential score significantly and, together with the online course, provide a well-rounded preparation for the test.

How to Use This Book?

Our goal is to simulate the entire experience of the actual test. By using this practice exam as suggested below, you will become familiar and comfortable with the environment experienced during the official exam. This alone can reduce stress and raise your potential score. Try to make an effort to follow the recommendations.

1. **Prepare a simulating environment:**

   - Free up at least 3 hours for taking the practice exam.

   - Prepare an empty desk - preferably, a college chair with an armrest.

   - Do not wear a watch. Make sure an analogue clock is avaiable.

   - Try to start the practice test at the same time your actual test is scheduled.

   - Prepare at least 5 sharpened number 2 pencils and an eraser.

   - Put away cellphones, water and food. Do not plan on bathroom breaks.

2. **Taking the practice exam:**

   - USE THE BOOK AS IF IT WERE YOUR EXAM BOOKLET. Use the scratch pages, make notes and draw sketches on these actual pages. This book is specifically designed for mimicking the actually test experience.

   - Print out the designated answer sheet for this exam. It can be obtained from the ETS website[1]. Mark your answers there and, later, grade your test only by looking at the answer sheet. Marking your answers correctly on the answer sheet is crucial and worth practicing.

   - Give yourself exactly 2 hours and 50 minutes, as it will be on the actual exam.

   - Grade yourself accurately[2] - one point on any right answer. Minus 0.25 point for every wrong answer.

3. **Additional notes:[3]**

   - It is important to use no.2 pencils of good quality. I recommend "Dixon®️ Ticonderoga". Note that the "pre-sharpened" are usually not sharpened enough. These can be found easily on Amazon®️, or any office supplies store.

---

[1]Currently on: https://www.ets.org/s/gre/pdf/practice_book_math.pdf, page 69

[2]Refer again to the above PDF file from ETS to estimate your 3 digit score (currently on page 67).

[3]Disclaimer: These recommendations are based on personal experience. GP Group does not have any relationship with these companies.

- Use a thin eraser. Otherwise, you might erase many answers on your answer sheet when attempting to correct one. I recommend "Paper Mate® Tuff Stuff Eraser Stick (SN64801)".

Useful Links

1. Our prep course site includes links to the different course modules, lectures and handouts. It also includes updates on new published exams, valuable lectures, and promo codes for discount on our course materials. There is an android app for delivering updates, handouts and daily questions. This can all be found on our course site.

   Enter   www.SubjectMath.com

   Enter   www.facebook.com/gpsubjectmath

2. The video lectures are published on Udemy. Look for our different "Subject Math" modules. Make sure to always check out www.SubjectMath.com for special discounts, before purchasing modules via Udemy.

   Enter   www.udemy.com

3. Our Android app is published on Google Play. Refer to www.SubjectMath.com to get it.

4. Official information about the test from ETS.

   Enter   www.ets.org/gre/subject/about/content/mathematics

5. Comments, corrections and ideas will be appreciated. Contact us at

   ⤳ subjectmath@gpgroupcompany.com

About the Author

GP Group is led by Gilad Pagi. Pagi graduated 1st in class while acquiring his B.S. in Math and B.S.+M.S. in Engineering. Pagi has more than 10 years of experience in teaching, including teaching positions in calculus and linear algebra courses, and both private and group tutoring. Pagi achieved a top score in the math subject exam (900), and currently pursuing his PhD in mathematics at the University of Michigan, Ann Arbor.

Acknowledgements

Special thanks to Caleb Springer, our exam "Debugger".

Before starting the practice exam, make sure to follow the instructions from the introduction of the book.

If you are ready, start the exam.

Good Luck!

PRACTICE TEST

Note:

- $\log(x)$ denotes the logarithm in the natural basis.

- $\mathbb{R}, \mathbb{C}, \mathbb{Q}, \mathbb{Z}, \mathbb{N}$ denote the real numbers, complex numbers, rational numbers, integers and natural (positive) integers respectively.

- Unless specified otherwise, I is the identity matrix.

- When A is a ring, $A[x]$ denotes the polynomials with coefficients in A.

- "Such that" may be abbreviated as "s.t.".

- $\arctan(x)$ and $\tan^{-1}(x)$ denote the inverse function of $\tan(x)$, and similarly for $\sin(x)$ and $\cos(x)$.

1).
$$\int_0^{3\frac{\pi}{2}} x \sin x \, dx = ?$$

(A) 0

(B) 1

(C) -1

(D) π

(E) $-\pi$

2). Let $f : (-5, \infty) \to \mathbb{R}$, $f(x) = x \log(x+5)$. Let r be the number of solutions to $f(x) = 0$, and let m be the number of local maximum points of f. What is r-m?

(A) 0

(B) 1

(C) 2

(D) 3

(E) None of the above.

USE FOR SCRATCH WORK

3). Let $A = \int_0^1 (x+1)$, $B = \int_0^1 \frac{1}{(x+1)}$, $C = \int_0^1 \frac{1}{(x+1)^2}$. Which of the following is true?

(A) $A \geq B \geq C$

(B) $C \geq B \geq A$

(C) $A \geq C \geq B$

(D) $C \geq A \geq B$

(E) $A \geq C \geq B$

---

4). Let $\tan^{-1}(t)$ denote the inverse of the tangent function. Let $f(x) = \frac{d}{dx}(2\tan^{-1}(2x))$. What is the value of $f(0)$?

(A) 1

(B) 4

(C) $\frac{1}{4}$

(D) $\frac{1}{2}$

(E) 2

USE FOR SCRATCH WORK

5). Let P be the plane in $\mathbb{R}^3$ containing the points $(1,0,0)$, $(0,0,1)$ and $(0,1,1)$. Let ℓ be the line containing the set $\{(2 - 3t, 2 - 3\sqrt{6}t, 2 - 3t) : t \in \mathbb{R}\}$. What is the angle between the plane P and the line ℓ?

(A) 60°

(B) 30°

(C) 45°

(D) 0

(E) None of the above.

6).

$$\lim_{n \to \infty} \sum_{k=1}^{n} \frac{\log(k)}{kn} = ?$$

(A) 0

(B) 1

(C) $\frac{e}{2}$

(D) $\frac{\log(2)}{2}$

(E) The limit does not exist.

USE FOR SCRATCH WORK

7). Let $A = \{a_n\}$ be a sequence of real numbers. Let $M = \sup A, m = \inf A$. Given that both M and m are real. Which of the following is true?

   I $\lim_{n \to \infty} a_n = M$ or $\lim_{n \to \infty} a_n = m$

   II For any $\epsilon > 0$, A contains at least one number in $(M - \epsilon, M]$

   III For any $\epsilon > 0$, A contains infinitely many numbers in $(M - \epsilon, M]$

   (A) I

   (B) II

   (C) III

   (D) II, III

   (E) I, II and III

8). Let A be a 3×3 real matrix. Which of the following is true?

   I A has at least one real eigenvalue.

   II There exists an antisymmetric matrix B such that $A - B$ is diagonalizable.

   III There exists $b \in \mathbb{R}$, such that $A - bI$ has a non trivial null space.

   (A) I

   (B) II

   (C) III

   (D) II, III

   (E) I, II and III

USE FOR SCRATCH WORK

9).

$$\sum_{n=0}^{\infty} (-1)^n \frac{(0.5)^{n+1}}{(2n+1)(2n+2)} = ?$$

(A) $\frac{1}{\sqrt{2}} \arctan\left(\frac{1}{\sqrt{2}}\right) - \frac{1}{2} \log\left(\frac{3}{2}\right)$

(B) $\frac{1}{2} \arctan\left(\frac{1}{2}\right) - \frac{1}{\sqrt{2}} \log\left(\frac{3}{\sqrt{2}}\right)$

(C) $\frac{1}{\sqrt{2}} \arctan\left(\frac{1}{2}\right) - \frac{1}{2} \log\left(\frac{3}{2}\right)$

(D) $\frac{1}{\sqrt{2}} \arctan\left(\frac{1}{2}\right) - \frac{1}{\sqrt{2}} \log\left(\frac{3}{2}\right)$

(E) $\frac{1}{\sqrt{2}} \arctan\left(\frac{1}{\sqrt{2}}\right) - \frac{1}{2} \log\left(\frac{3}{\sqrt{2}}\right)$

10). Let G be the graph of $y^2 - 4x^2 = 1$. Let T_1 be the tangent line to G at $(0,1)$, T_2 be the tangent line to G at $(-1, \sqrt{5})$, T_3 be the tangent line to G at $(-1, -\sqrt{5})$, and T_4 be the tangent line to G at $(-2, \sqrt{17})$. Denote m_i to be the slope of T_i, where $i = 1, 2, 3, 4$. Which of the following is true?

(A) $m_1 \leq m_2 \leq m_3 \leq m_4$

(B) $m_4 \leq m_3 \leq m_2 \leq m_1$

(C) $m_4 \leq m_2 \leq m_1, m_2 = m_3$

(D) $m_4 \leq m_2 \leq m_1 \leq m_3$

(E) None of the above

USE FOR SCRATCH WORK

11). Let C be a circle in the first quadrant, such that C is tangent to both axes and the point $p = (\frac{1}{\sqrt{2}}, \frac{1}{\sqrt{2}})$ is the point on C most distant from the origin. What is the radius of C?

(A) $\frac{1}{2}$

(B) $\frac{1}{2\sqrt{2}}$

(C) $\sqrt{2} - 1$

(D) $\sqrt{2} + 1$

(E) Such circle does not exists.

12). Let C be a curve in the xy-plane with the following parametrization: $x = 3\cos t$, $y = 2\sin t, t \in [0, \frac{3\pi}{2}]$. Let L be its length. Which of the following is true?

(A) $0 \le \frac{L}{\pi} < 1$

(B) $1 \le \frac{L}{\pi} < 2$

(C) $2 \le \frac{L}{\pi} < 3$

(D) $3 \le \frac{L}{\pi} < 5$

(E) $5 \le \frac{L}{\pi} < 7$

USE FOR SCRATCH WORK

13). Let G be a group of order 34. Which of the following is true?

    I G must be abelian.

    II G has a subgroup of order 17.

    III G must have at least one proper non-trivial normal subgroup.

(A) I

(B) II

(C) I, II

(D) II, III

(E) I, II and III

---

14). The following is a sketch of the graph of $f(x) = x^3 + ax^2 + bx + c$.

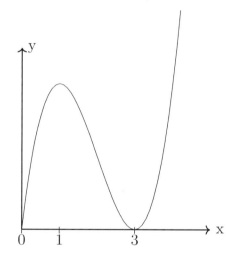

What is the value of $a + b + c$?

(A) 0

(B) 3

(C) 6

(D) 9

(E) 12

USE FOR SCRATCH WORK

15). Which of the following most closely represents the graph of a solution to the differential equation $\frac{dy}{dx} = xy$?

(A)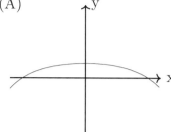

(B)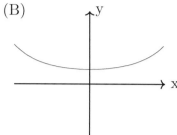

(C)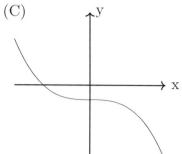

(D)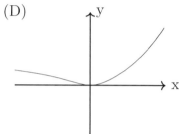

(E)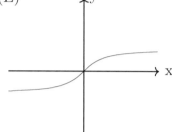

USE FOR SCRATCH WORK

16). What is the minimal distance between the plane $x + 2y + 3z = 28$ and the sphere $x^2 + y^2 + z^2 = 2$?

(A) $\sqrt{14}$

(B) $2\sqrt{14}$

(C) $2\sqrt{14} - 1$

(D) $2\sqrt{14} - 2$

(E) $2\sqrt{14} - \sqrt{2}$

17). Let T be an isosceles triangle with base and height of length 2. Let $ABCD$ be a (convex) rectangle where two adjacent vertices, A and B, are located on T's base, C is located on one of T's legs and D in located on the other leg. What is the maximum possible area of $ABCD$?

(A) 0

(B) $\frac{1}{2}$

(C) 1

(D) 2

(E) None of the above.

USE FOR SCRATCH WORK

18). Let $F(x) : (0, \infty) \to \mathbb{R}$ defined as follows:

$$F(x) = \int_x^{x^2} \frac{\log(t)}{t} dt.$$

Which of the following is true?

    I $\lim_{x \to \infty} F(x)$ exists and is bounded.

    II $\lim_{x \to \infty} \frac{dF}{dx}$ exists and is bounded.

    III $\lim_{x \to 0^+} \frac{dF}{dx}$ exists and is bounded.

(A) I

(B) II

(C) III

(D) I, II

(E) I, II and III

---

19). Let $g(x) = |x - 1| + |x - 2|$.

$$\int_0^3 g(x) dx = ?$$

(A) 0

(B) 1

(C) 2

(D) 2.5

(E) 5

USE FOR SCRATCH WORK

20). Which of the following is FALSE?

    I A continuous map is an open map.

    II An open map is a continuous map.

    III A closed map is an open map.

(A) I

(B) II

(C) III

(D) I, II

(E) I, II and III

---

21). Let (a_n) be a sequence or real numbers. Which of the following is true?

(A) If $\lim_{n \to \infty}(a_n) = 1$ then it has a subsequence which converges to 0.

(B) If $\lim_{n \to \infty}(a_n) = 0$ then $\sum_{n=0}^{\infty} a_n$ converges.

(C) If $\lim_{n \to \infty}(a_n) = 0$ then there exists a subsequence (a_{n_k}) of (a_n) such that $\sum_{k=0}^{\infty} a_{n_k}$ is convergent.

(D) $\{a_n\}$ is a closed set of $\mathbb{R}$.

(E) None of the above is a true statement.

USE FOR SCRATCH WORK

22). A k-row is composed of a row of k (ordered) numbers $(x_1, x_2, ..., x_k)$ where the number in each slot is an integer from 1 to 10, and any number appears only once (i.e. there are no 11-row). A well-ordered k-row is a k-row in which for $i > j, x_i > x_j$. (i.e. (1,2,4) is a well ordered 3-row, but (1,4,2) is not). Consider a process in which a 7-row is chosen randomly. What is the probability to choose a well ordered 7-row?

(A) $\frac{1}{7!}$

(B) $\frac{1}{7!3!}$

(C) $\frac{1}{\binom{10}{7}}$

(D) $\frac{1}{\binom{10}{3}}$

(E) None of the above.

23). Which of the following is true for a function $f(x) : (0, \infty) \to \mathbb{R} \setminus \{0\}$?

I $\lim_{x \to \infty} f = 0$ implies that $\frac{1}{f}$ is not bounded.

II $\lim_{x \to \infty} \frac{1}{f} = \infty$ implies that f is bounded.

III $\frac{1}{f}$ is unbounded in every interval containing $x = 3$ implies that $\lim_{x \to 3} f = 0$

(A) I

(B) II

(C) III

(D) I, III

(E) I, II and III

USE FOR SCRATCH WORK

24). Let $f : X \to X$ be a function for a set X to X. Let $A \subseteq X$. Denote $B = f(f^{-1}(A))$, $C = f^{-1}(f(A))$. Which of the following is true?

(A) $A \subseteq B \subseteq C$

(B) $A \subseteq C \subseteq B$

(C) $B \subseteq A \subseteq C$

(D) $C \subseteq A \subseteq B$

(E) $B \subseteq C \subseteq A$

25). What is the volume of the solid formed by revolving about the x-axis the region in the first quadrant of the xy-plane bounded by the coordinate axes and the the graph of $y = \frac{1}{x+1}$?

(A) π

(B) π^2

(C) 2π

(D) $\frac{\pi}{2}$

(E) ∞

USE FOR SCRATCH WORK

26). Let A, B, C, D be 4×4 real matrices. Which of the following cannot be a true statement?

(A) The null space of BC is properly contained in the null space of ABC

(B) The null space of BC is properly contained in the null space of C

(C) The column space of AB is properly contained in the column space of CD

(D) The column space of BCD is properly contained in the column space of BC

(E) All of the above may be true.

27). Denote $\mathbb{R}[x]$ as the set of polynomials with real coefficients. Which of the following is a vector space?

I $\{A \in \mathbb{C}^{n \times n} \,|\, A = -A^T\}$ no identity (69)

II $\{p(x) \in \mathbb{R}[x] \,|\, p(x) = 0, \text{ or } \deg(p(x)) \geq 2\}$.

III $\{x \in \mathbb{C}^n \,|\, \exists \lambda \in \mathbb{C} \, s.t. \, Ax = \lambda x\}$, where $A \in \mathbb{C}^{n \times n}$ is fixed.

(A) I

(B) II

(C) III

(D) I, III

(E) I, II and III

USE FOR SCRATCH WORK

28). Let V be a vector space and $u, v, w \in V$ be vectors. Which of the following is true?

(A) If $\{u, v\}$ is a linearly dependent set and $\{u, w\}$ is a linearly dependent set, then $\{v, w\}$ is a linearly dependent set.

(B) If $\{u, v, w\}$ is a linearly dependent set then w is a linear combination of u and v.

(C) If $V = \text{Span}(\{u, v\})$ then V is of dimension 2.

(D) If V is of dimension 4, then exist a forth vector $x \in V$, such that $V = \text{Span}(\{u, v, w, x\})$

(E) None of the above.

---

29). Let n be an integer bigger than 2 and let $f(x, y) = 3x^2 - 6xy + y^n$. Which of the following is true?

(A) f has no critical points for any value of n.

(B) Exists n such that f has exactly one critical point.

(C) Exists n such that f has a local minimum at the origin.

(D) Exists n such that f has a local minimum at point different than the origin.

(E) None of the above.

USE FOR SCRATCH WORK

30). Let P_1 be the plane $x + 2y - 4z = 0$. Let P_2 be the plane $10x + by + cz = 0$. Given that P_1 and P_2 are perpendicular to each other, and that their intersection is contained in the yz-plane. What is the value of $b + c$?

(A) 0

(B) 1

(C) -1

(D) 2

(E) None of the above.

31). Let $f(z) = f(x + iy) = u(x, y) + iv(x, y)$ be an entire complex function (analytic on the entire complex plane) and let $f'(z)$ denote its derivative. Given $f(0) = 0, f'(z) = 2z$. What is the value of $u(1, 2)$?

(A) 0

(B) 2

(C) -2

(D) 3

(E) -3

USE FOR SCRATCH WORK

32). Let G be the set of roots of $f(x) = x^7 - 1$ in the complex plane $\mathbb{C}$. Which of the following is FALSE?

(A) G is a multiplicative group with the multiplication operation from $\mathbb{C}$.

(B) G is an abelian group.

(C) G has a proper normal subgroup.

(D) $\prod_{g \in G} g = 1$.

(E) All the statements above are true.

USE FOR SCRATCH WORK

33). The following figure contains sketches of the graphs of $f(x) = x$, $g(x) = \frac{1}{x}$, $h(x) = \frac{x^2}{2}$:

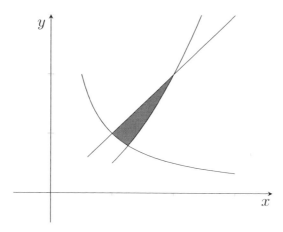

If A is the area of the gray region, what is the value of $3 - 6A$?

(A) $\log(2)$

(B) $\log(4)$

(C) $-\log(2)$

(D) $-\log(4)$

(E) None of the above.

$2\ln(2)$

USE FOR SCRATCH WORK

34).

$$\lim_{N \to \infty} \sum_{k=1}^{N} \frac{\sin\left(\frac{k}{N}\pi\right)}{N} = ?$$

(A) 1

(B) 2

(C) $\frac{1}{\pi}$

(D) $\frac{2}{\pi}$

(E) $\frac{1}{2\pi}$

35). Observe the following function:

$$f(x) = \begin{cases} 0 & x \in \mathbb{Q} \\ 1 & \text{Otherwise} \end{cases}$$

What is the cardinality of the set of points of discontinuity of $f(x)$?

(A) 0

(B) 1

(C) countably infinite.

(D) uncountably infinite.

(E) None of the above.

USE FOR SCRATCH WORK

36). Let A be a matrix with characteristic polynomial $c_A(x) = x^3 - 6x^2 + 11x - 6$. What is the trace of A^{-1}?

(A) $\frac{1}{6}$

(B) $\frac{5}{6}$

(C) $\frac{11}{6}$

(D) 6

(E) None of the above

37). Let $A = \begin{pmatrix} 0 & 1 \\ 1 & 0 \end{pmatrix}$, and define $T : \mathbb{R}^{2\times 2} \to \mathbb{R}^{2\times 2}$ to be the linear transformation $T(X) = XA - AX$. Let U be the subspace of $\mathbb{R}^{2\times 2}$ spanned by $\left\{ \begin{pmatrix} 1 & 0 \\ 0 & 0 \end{pmatrix}, \begin{pmatrix} 0 & 1 \\ 0 & 0 \end{pmatrix} \right\}$. Let L be the restriction of T to U. What is the value of $\dim \operatorname{Im} T - \dim \operatorname{Im} L$?

(A) 0

(B) 1

(C) 2

(D) 3

(E) 4

USE FOR SCRATCH WORK

38). Let $\mathbf{F}$ be a vector field in $\mathbb{R}^3$, $\mathbf{F} = (x^2, -xy, (y+1)z)$. What is the work done by $\mathbf{F}$ on a particle that moves a along the path $(-t, t, \log(t+1))$ between time $t = 0$ and time $t = 1$?

(A) $\log(2)$

(B) $\log(4)$

(C) $-\log(2)$

(D) $-\log(4)$

(E) $\log(4) - 1$

39). What is the global maximum of the function $f(x) = x^2 + y$ in the region $x^2 + y^2 \leq 1$?

(A) 0

(B) 0.5

(C) 1

(D) 1.25

(E) 1.5

USE FOR SCRATCH WORK

40). Let $\lfloor x \rfloor$ denote the greatest integer not exceeding x.

$$\sum_{k=0}^{\infty} (-1)^{\lfloor \frac{k}{2} \rfloor} \frac{\pi^k}{k!} = ?$$

(A) 0

(B) 1

(C) -1

(D) π.

(E) $-\pi$.

41). Let $A = \{0\} \cup \{1\} \cup (2,4)$, $B = \{0\} \cup \{3\}$, be subsets of $\mathbb{R}$ with the standard topology, denoted as $(\mathbb{R}, T)$. Denote (A, T_A) the subset A with the subset topology. Which of the following is true? (Recall that y an isolated point of Y in (X, T_X) if there exists a neighborhood U of y in (X, T_X) such that $U \cap Y = \{y\}$.)

I A, as a subset of $(\mathbb{R}, T)$, has 2 isolated points.

II B, as a subset of (A, T_A), has 2 isolated points.

III B, as a subset of (A, T_A), has 1 interior point.

(A) I

(B) II

(C) III

(D) I, II

(E) I, II and III

USE FOR SCRATCH WORK

42). Let A be $\begin{pmatrix} 1 & k & 2(1-k) \\ 0 & k & 1-k \\ 1 & k & 1-k \end{pmatrix}$, where $k \in \mathbb{C}$. Let b be $\begin{pmatrix} 0 \\ 0 \\ 1 \end{pmatrix}$. For which value of k, the system $Ax = b$ has infinitely many solutions?

(A) 1

(B) 0

(C) -1

(D) 0 and 1

(E) None of the above.

USE FOR SCRATCH WORK

43). Which of the following is true for a group G?

     I If G is of order 4, then G is abelian.

    II If G is of order 6, then G is abelian.

  III If all the subgroups of G are normal, then G is abelian.

(A) I

(B) II, II

(C) I,II

(D) I,III

(E) I,II and III

USE FOR SCRATCH WORK

44). Denote $d = \gcd(330, 105), m = \text{lcm}(330, 105)$. What is $\gcd(d, m)$?

(A) 5

(B) 3

(C) 15

(D) 330

(E) 462

45). Which of the following is the best approximation of $(8008)^{\frac{2}{3}}$?

(A) 330

(B) 400

(C) 460

(D) 550

(E) 1000

USE FOR SCRATCH WORK

46). Consider the the following statement:"Let a_n and b_n be sequences of positive real numbers. If $\lim_{n\to\infty}\frac{a_n}{b_n}=1$, then $\lim_{n\to\infty}(a_n-b_n)=0$."

Consider the following attempt to prove this statement:

(1) $\lim_{n\to\infty}\frac{a_n}{b_n}=1$, then for any $\epsilon>0$ exists N such that for $n>N$:

$$\left|\frac{a_n}{b_n}-1\right|\leq\epsilon$$

(2) Thus for $n>N$:
$$|a_n-b_n|\leq\epsilon b_n$$

(3) Since the latter is true for an arbitrary ϵ, then when $N\to\infty$, $|a_n-b_n|\to0$.

(4) Therefore, $\lim_{n\to\infty}(a_n-b_n)=0$

Which of the following is True?

(A) The argument is valid

(B) The inference in step (1) is wrong.

(C) The inference in step (2) is wrong.

(D) The inference in step (3) is wrong.

(E) The inference in step (4) is wrong.

USE FOR SCRATCH WORK

47). Let $C = \{z : |z| = 1\}$ by the unit circle in the complex plane, oriented counter-clockwise.

$$\oint_C \frac{\sin(z) + \cos(z)}{z} \, dz = ?$$

(A) $2\pi i$

(B) $4\pi i$

(C) πi

(D) $\frac{1}{2}\pi i$

(E) None of the above.

USE FOR SCRATCH WORK

48). By definition, a function $f(x) : \mathbb{R} \to \mathbb{R}$ is uniformly continuous if for any $\epsilon > 0$, exists $\delta > 0$ such that $|x - y| < \delta$ implies $|f(x) - f(y)| < \epsilon$. Which of the following statements is the negation to the statement of the latter definition?

(A) $\exists \epsilon > 0$ for which there exists $\delta > 0$ such that $|x-y| < \delta$ implies $|f(x)-f(y)| \geq \epsilon$

(B) $\forall \epsilon > 0$, $\exists \delta > 0$ such that $|x - y| < \delta$ implies $|f(x) - f(y)| \geq \epsilon$

(C) $\exists \epsilon > 0$ for which $\forall \delta > 0$, $|x - y| < \delta$ implies $|f(x) - f(y)| \geq \epsilon$

(D) $\exists \epsilon > 0$ for which $\forall \delta > 0$ there exists $x, y \in \mathbb{R}$ with $|x-y| < \delta$ and $|f(x)-f(y)| \geq \epsilon$

(E) None of the above.

---

49). Observe the following power series:

$$\sum_{k=0}^{\infty} \frac{(kx)^k}{k!}$$

What is the radius of convergence?

(A) 0

(B) ∞

(C) e

(D) $\frac{1}{e}$

(E) None of the above.

USE FOR SCRATCH WORK

50). Let $\mathbb{N}$ be the positive integers and define a function $f : \mathbb{N} \times \mathbb{N} \to \{0, 1, ..., 9\} \times \{0, 1, ..., 4\}$ by

$$f(a, b) = (a \mod 10, (a + b) \mod 5).$$

What is the minimum number m, such that for any $A \subset \mathbb{N} \times \mathbb{N}$ with cardinality m there exists $(x, y) \in \mathbb{N} \times \mathbb{N}$ with $|f^{-1}(x, y) \cap A| > 4$?

(A) 36

(B) 51

(C) 145

(D) 151

(E) 201

51). Which of the following polynomials has a root in neither the field of 3 elements $\mathbb{F}_3$, nor the field of 5 elements $\mathbb{F}_5$?

(A) $x^2 + x + 1$

(B) $x + 1$

(C) $x^2 + 1$

(D) $x^2 + 2x + 1$

(E) $x^2 + x + 2$

USE FOR SCRATCH WORK

52). Let C be an ellipse in the xy-plane passing through $(2,0), (0,3)$ and $(-2,0)$, oriented counterclockwise.

$$\oint_C (2x\sin(y) - y)dx + (x^2\cos(y) + x)dy = ?$$

(A) 0

(B) 6π

(C) 12π

(D) 18π

(E) 24π

---

53). Let A be a matrix with characteristic polynomial $c_A(x) = x^3 + x^2 + 1$. What is the trace of the inverse of A^2?

$\left[\text{Cayley - Hamilton} \right]$

(A) 0

(B) 1

(C) -1

(D) 2

(E) -2

USE FOR SCRATCH WORK

54). Let $A \in \mathbb{C}^{n \times n}$ be a square complex matrix. Which of the following DO NOT imply that A is invertible?

(A) $Ax = 0$ has a single solution, for $x \in \mathbb{C}^n$.

(B) The columns of A are linearly independent.

(C) The characteristic polynomial of A is $x^5 + x^3 + x + 15$.

(D) A is diagonalizable, and its trace is 3.

(E) None of the above.

55). Which of the following imply that a measurable set $A \subset \mathbb{R}$ has a positive measure?

(A) A is uncountable.

(B) A is dense.

(C) A is compact.

(D) $\mathbb{R} \setminus A$ is compact

(E) None of the above.

USE FOR SCRATCH WORK

56). Let a, b, c be the eigenvalues of:

$$A = \begin{bmatrix} i & 1 & 1 \\ 1 & i & 1 \\ 1 & 1 & i \end{bmatrix}$$

What is the value of $(a+1)(b+1)(c+1)$?

(A) 0

(B) 1

(C) i

(D) $1+i$

(E) None of the above

---

57). Let $f : \mathbb{R} \to \mathbb{R}$ be a continuous function. Which of the following is true?

    I If $K \subset \mathbb{R}$ is compact, then $f(K)$ is compact.

    II If $K \subset \mathbb{R}$ is compact, then $f^{-1}(K)$ is compact.

    III If $K \subset \mathbb{R}$ is connected, then $f(K)$ is connected.

(A) I

(B) I, II

(C) II, III

(D) I, III

(E) I, II and III

USE FOR SCRATCH WORK

58). Consider the sequence of real functions: $f_n(x) = \sin^n(x)$. Consider the following intervals: $I_1 = (-4, \frac{\pi}{2}], I_2 = (-\frac{\pi}{2}, \frac{\pi}{2}), I_3 = (-\frac{\pi}{2}, \frac{\pi}{2}], I_4 = (-\frac{3}{2}, \frac{3}{2}]$. We say that an interval I_j is "bigger" than I_k if $I_j \supset I_k$. Which of those intervals is the biggest interval in which f_n converges uniformly?

(A) I_1

(B) I_2

(C) I_3

(D) I_4

(E) None of the above.

59). Let M be a module over the ring $\mathbb{Z}_5$, which denotes the set $\{0, 1, 2, 3, 4\}$ with addition and multiplication modulo 5. Which of the following must be true?

(A) M is finite.

(B) M is a vector space.

(C) If M is a ring, the multiplication in M is commutative.

(D) M must be isomorphic to $\mathbb{Z}_5$.

(E) None of the above must be true.

USE FOR SCRATCH WORK

60). A **complete** graph of n vertices, denoted K_n, is an undirected graph with n vertices and a unique edge connecting each pair of vertices. An **Eulerian** cycle is a cycle passing through every edge exactly once. Let $E(n)$ denote the number of edges in K_n. For which n, K_n has an Eulerian cycle and $E(n)$ is divisible by 5?

(A) 4

(B) 5

(C) 6

(D) 10

(E) 13

For any ring A, let $A[x]$ denote the polynomials with coefficients in A. Let A_k, where $k \in \mathbb{R}$, denote the set $\{k\} \times A$ (cartesian product). Let $\mathbb{Q}$ denote the rational numbers, and $\mathbb{N}$ - the natural numbers. Which of the following sets has a different cardinality than the others?

61). (A) $\bigcup_{k \in \mathbb{Q}} \mathbb{Q}_k$

(B) $\mathbb{Q}[x]$.

(C) $\mathbb{Z}[x]$

(D) $\mathbb{Q} \times \mathbb{Q}$ (cartesian product).

(E) The set of all functions from $\mathbb{N}$ to $\mathbb{N}$.

USE FOR SCRATCH WORK

62).

$$i \tan(i) = ?$$

(A) $\frac{1-e^2}{1+e^2}$

(B) $\frac{1+e^2}{1-e^2}$

(C) $\frac{1-e}{1+e}$

(D) $\frac{1+e}{1-e}$

(E) $\frac{1+e^2}{1-e^2} i$

63). How many abelian groups of order 540 do not have a subgroup isomorphic to the Klein four-group? (Only count the groups up to isomorphism.)

(A) 2

(B) 3

(C) 4

(D) 5

(E) 6

USE FOR SCRATCH WORK

64). Let $A = \{a_n\}_{n \in \mathbb{N}}$ be a sequence of real numbers. Which of the following is true?

    (A) A is closed.

    (B) A is compact.

    (C) If A is bounded, then A is compact.

    (D) If $\lim_{n \to \infty} a_n$ does not exist, then A is closed.

    (E) If $\lim_{n \to \infty} a_n = L$ And $L \in A$ then A is compact.

---

65). Let $\mathbb{R}$ denote the real line with the standard topology. Let X denote the real line with the following topology: $U \subset X$ is open $\iff$ $X \setminus U$ is finite or $U = \emptyset$. Which of the following is true?

    I Exists a proper set $A \neq \emptyset$ which is compact in both X and $\mathbb{R}$.

    II Exists a proper set $A \neq \emptyset$ which is open in both X and $\mathbb{R}$.

    III Every subset of X is compact.

    (A) I

    (B) II

    (C) III

    (D) I, III

    (E) I, II and III

USE FOR SCRATCH WORK

66). Let $f(x) : \mathbb{R} \to \mathbb{R}$ be a continuous function. Given $f(-1) = 1$, $f(1) = 3$. Which of the following is the weakest requirement to ensure the existence of $c \in (-1, 1)$ with $\frac{df}{dx}(c) = 1$?

(A) f is differentiable on $[-1, 1]$ and $\frac{df}{dx}$ is bounded there.

(B) f is differentiable on $[-1, 1]$.

(C) f is differentiable on $(-1, 1)$ and $\frac{df}{dx}$ is continuous there.

(D) f is differentiable on $(-1, 1)$.

(E) f is differentiable on $\mathbb{R}$ and $\frac{df}{dx}$ is continuous.

END OF PRACTICE TEST

Answers and Solutions

1).
$$\int_0^{3\frac{\pi}{2}} x \sin x \, dx = ?$$

(A) 0

(B) 1

(C) -1

(D) π

(E) $-\pi$

The answer is (C). We use Integration by parts:

$$\int_0^{3\frac{\pi}{2}} x \sin x \, dx = x(-\cos x)\Big|_0^{3\frac{\pi}{2}} - \int_0^{3\frac{\pi}{2}} (-\cos x)dx = 0 + \int_0^{3\frac{\pi}{2}} \cos x \, dx = -1.$$

Notice that $\cos(x)$ is zero for $x = \frac{\pi}{2}, 3\frac{\pi}{2}$, and we don't have to calculate the last integral if we use the following trick, as taught in the Advanced Module:

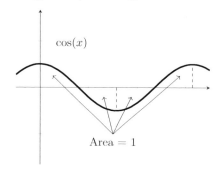

2). Let $f : (-5, \infty) \to \mathbb{R}$, $f(x) = x \log(x + 5)$. Let r be the number of solutions to $f(x) = 0$, and let m be the number of local maximum points of f. What is r-m?

(A) 0

(B) 1

(C) 2

(D) 3

(E) None of the above.

Answer is (C). Even tough f is not a polynomial we use the term "roots" for solutions of $f(x) = 0$. f has roots when one of the terms is 0, that is when $x = 0, -4$. $f' = \log(x + 5) + \frac{x}{x+5}$, and $f'' = \frac{1}{x+5} + \frac{5}{(x+5)^2}$ by the formula for $\left(\frac{f}{g}\right)'$. Since the second derivative is always positive, any critical point, if exists, would be a local minimum point. So $r = 2, m = 0$. As a reference, below is the graph of the function.

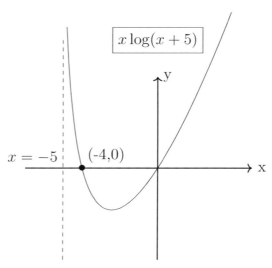

$x \log(x + 5)$

$x = -5$ (-4,0)

3). Let $A = \int_0^1 (x + 1)$, $B = \int_0^1 \frac{1}{(x+1)}$, $C = \int_0^1 \frac{1}{(x+1)^2}$. Which of the following is true?

(A) $A \geq B \geq C$

(B) $C \geq B \geq A$

(C) $A \geq C \geq B$

(D) $C \geq A \geq B$

(E) $A \geq C \geq B$

Answer is (A). It is possible to compute the integrals (which are 1.5, log(2), and 0.5, respectively), but this is not time efficient. Recall that the function $g(t) = c^t$ is non-decreasing as long as $c \geq 1$. So for $f(x) = (x + 1)$, and a fixed x, applying f^t for $t = -2, -1, 1$ makes the value to increase accordingly. As a result, being all positive, the values of the integrals agree with (A).

This problem is typical for the exam. It is possible to compute everything and get the right answer while spending 3-4 minutes on the solution. However a better "general" approach, and confidence, lead to the correct answer in 1 minute.

4). Let $\tan^{-1}(t)$ denote the inverse of the tangent function. Let $f(x) = \frac{d}{dx}(2 \tan^{-1}(2x))$. What is the value of $f(0)$?

(A) 1

(B) 4

(C) $\frac{1}{4}$

(D) $\frac{1}{2}$

(E) 2

Answer is (B). Take the derivative and mind the additional 2 due to the composition. $f(x) = 2\frac{1}{1+(2x)^2} \cdot 2$.

5). Let P be the plane in $\mathbb{R}^3$ containing the points $(1,0,0)$, $(0,0,1)$ and $(0,1,1)$. Let ℓ be the line containing the set $\{(2 - 3t, 2 - 3\sqrt{6}t, 2 - 3t) : t \in \mathbb{R}\}$. What is the angle between the plane P and the line ℓ?

(A) 60°

(B) 30°

(C) 45°

(D) 0

(E) None of the above.

Answer is (B). By subtracting the point $(0,0,1)$ from the others, the plane contains the vectors $(1, 0, -1), (0, 1, 0)$ so it is clear that $(1, 0, 1)$ is the normal for the plane. The line is directed along the vector $(1, \sqrt{6}, 1)$. The angle α between the normal and the vector would complement the angle in question, θ to 90°. By computing the dot product, $\cos(\alpha) = \frac{(1,0,1)\cdot(1,\sqrt{6},1)}{\sqrt{2}\cdot\sqrt{8}} = \frac{1}{2} \Rightarrow \alpha = 60^o \Rightarrow \theta = 30^o$.

6).

$$\lim_{n\to\infty} \sum_{k=1}^{n} \frac{\log(k)}{kn} =?$$

(A) 0

(B) 1

(C) $\frac{e}{2}$

(D) $\frac{\log(2)}{2}$

(E) The limit does not exist.

The answer is (A). It is tempting to try solving that using Riemann sums, but that is the wrong approach here. This is the arithmetic mean of the sequence $\frac{\log(n)}{n}$ which aspires to zero, and so is its arithmetic mean.

7). Let $A = \{a_n\}$ be a sequence of real numbers. Let $M = \sup A, m = \inf A$. Given that both M and m are real. Which of the following is true?

  I $\lim_{n\to\infty} a_n = M$ or $\lim_{n\to\infty} a_n = m$

  II For any $\epsilon > 0$, A contains at least one number in $(M - \epsilon, M]$

  III For any $\epsilon > 0$, A contains infinitely many numbers in $(M - \epsilon, M]$

(A) I

(B) II

(C) III

(D) II, III

(E) I, II and III

Answer is (B). $a_n = (-1)^n$ is a counter example for I, since it does not have a limit at all. It also serves as a counter example for III since A contains only 2 numbers so the statement at III cannot be true. Finally, II is true since otherwise, suppose exists ϵ such that A does not contain any number in $(M - \epsilon, M]$. Then the supremum of A is at most $M - \epsilon$, which contradicts the assumption.

8). Let A be a 3×3 real matrix. Which of the following is true?

   I A has at least one real eigenvalue.

   II There exists an antisymmetric matrix B such that $A - B$ is diagonalizable.

   III There exists $b \in \mathbb{R}$, such that $A - bI$ has a non trivial null space.

(A) I

(B) II

(C) III

(D) II, III

(E) I, II and III

Answer is (E). I and III are basically the same statement. If A has an eigenvalue λ, then the null space of $A - \lambda I$ is the eigenspace of λ so it must be non trivial, and vice versa. Since A is real and 3 by 3, its characteristic polynomial is of degree 3 with real coefficients. Any odd degree polynomial with real coefficients has at least one real root, so the characteristic polynomial has a real root. So both I and III are true. For any matrix $A = \frac{A+A^T}{2} + \frac{A-A^T}{2}$, that is a sum of a symmetric and an antisymmetric matrices. Denote $B = \frac{A-A^T}{2}$ and we get that $A - B$ is a real symmetric matrix and thus diagonalizable.

9).

$$\sum_{k=0}^{\infty} (-1)^n \frac{(0.5)^{n+1}}{(2n+1)(2n+2)} = ?$$

(A) $\frac{1}{\sqrt{2}} \arctan\left(\frac{1}{\sqrt{2}}\right) - \frac{1}{2} \log\left(\frac{3}{2}\right)$

(B) $\frac{1}{2} \arctan\left(\frac{1}{2}\right) - \frac{1}{\sqrt{2}} \log\left(\frac{3}{\sqrt{2}}\right)$

(C) $\frac{1}{\sqrt{2}}\arctan\left(\frac{1}{2}\right) - \frac{1}{2}\log\left(\frac{3}{2}\right)$

(D) $\frac{1}{\sqrt{2}}\arctan\left(\frac{1}{2}\right) - \frac{1}{\sqrt{2}}\log\left(\frac{3}{2}\right)$

(E) $\frac{1}{\sqrt{2}}\arctan\left(\frac{1}{\sqrt{2}}\right) - \frac{1}{2}\log\left(\frac{3}{\sqrt{2}}\right)$

Answer is (A). The series in question resembles the series of

$$\arctan(x) = \sum_{k=0}^{\infty}(-1)^n \frac{x^{2n+1}}{(2n+1)}.$$

Let us integrate:

$$\int \arctan(x) = \sum_{k=0}^{\infty}(-1)^n \frac{x^{2n+2}}{(2n+1)(2n+2)} = \sum_{k=0}^{\infty}(-1)^n \frac{(x^2)^{n+1}}{(2n+1)(2n+2)} = F(x).$$

So we get the desired value by plugging in $x = 1/\sqrt{2}$. On the other hand, we use integration by parts to get:

$$\int \arctan(x)dx = \int 1 \cdot \arctan(x)dx =$$

$$= x\arctan(x) - \int \frac{x}{1+x^2}$$

$$= x\arctan(x) - \frac{1}{2}\log(1+x^2) + C = F(x)$$

Evaluate $F(0)$ to get $C = 0$. Finally plug in $x = 1/\sqrt{2}$ to get the answer.

10). Let G be the graph of $y^2 - 4x^2 = 1$. Let T_1 be the tangent line to G at $(0,1)$, T_2 be the tangent line to G at $(-1, \sqrt{5})$, T_3 be the tangent line to G at $(-1, -\sqrt{5})$, and T_4 be the tangent line to G at $(-2, \sqrt{17})$. Denote m_i to be the slope of T_i, where $i = 1, 2, 3, 4$. Which of the following is true?

(A) $m_1 \le m_2 \le m_3 \le m_4$

(B) $m_4 \le m_3 \le m_2 \le m_1$

(C) $m_4 \le m_2 \le m_1, m_2 = m_3$

(D) $m_4 \le m_2 \le m_1 \le m_3$

(E) None of the above

Answer is (D). This it a sketch of the hyperbola:

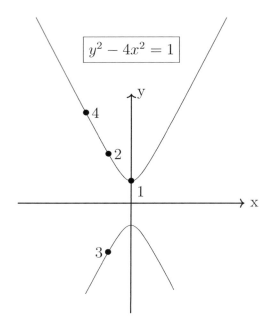

m_1 is 0, m_3 is the only positive slope out of the four (it is equal to m_2 by absolute value but not as slopes). m_4 is closer to the asymptotic slope of -2 then m_2 (recall that the slope in an hyperbola changes monotonically). So all those considerations leads to $m_4 < m_2 < m_1 < m_3$.

11). Let C be a circle in the first quadrant, such that C is tangent to both axes and the point $p = (\frac{1}{\sqrt{2}}, \frac{1}{\sqrt{2}})$ is the point on C most distant from the origin. What is the radius of C?

(A) $\frac{1}{2}$

(B) $\frac{1}{2\sqrt{2}}$

(C) $\sqrt{2} - 1$

(D) $\sqrt{2} + 1$

(E) Such circle does not exists.

Answer is (C). Observe the sketch below. C is tangent to the axes implies that the center is on the line $y = x$ and the equation for C is $(x - r)^2 + (y - r)^2 = r^2$. Note: r must be smaller than $\frac{1}{\sqrt{2}} \approx 1.4/2 \approx 0.7$, so eliminate (B) and (D). For any point on $x = y$ exists a circle consisting that point as the most distant from the origin so eliminate (E). Plug-in p to get the answer.

Alternative solution - by using the dotted right triangle we can infer the following.

$$\frac{1}{\sqrt{2}} = r + r \cdot \frac{1}{\sqrt{2}} \Rightarrow r = \frac{\frac{1}{\sqrt{2}}}{1 + \frac{1}{\sqrt{2}}} = \frac{\frac{1}{\sqrt{2}}}{1 + \frac{1}{\sqrt{2}}} \cdot \frac{1 - \frac{1}{\sqrt{2}}}{1 - \frac{1}{\sqrt{2}}} = \frac{\frac{1}{\sqrt{2}} - \frac{1}{2}}{\frac{1}{2}} = \sqrt{2} - 1$$

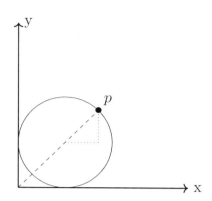

12). Let C be a curve in the xy-plane with the following parametrization: $x = 3\cos t$, $y = 2\sin t, t \in [0, \frac{3\pi}{2}]$. Let L be its length. Which of the following is true?

(A) $0 \leq \frac{L}{\pi} < 1$

(B) $1 \leq \frac{L}{\pi} < 2$

(C) $2 \leq \frac{L}{\pi} < 3$

(D) $3 \leq \frac{L}{\pi} < 5$

(E) $5 \leq \frac{L}{\pi} < 7$

Answer is (D). The parametrization implies that C is three quarters of an ellipse with radii 2, 3. So its length is bounded below by three quarters of the perimeter of a circle of radius 2, and bounded above by three quarters of the perimeter of a circle of radius 3. So $L > \frac{3}{4}2\pi \cdot 2 = 3\pi$, $L < \frac{3}{4}2\pi \cdot 3 = 4.5\pi$.

13). Let G be a group of order 34. Which of the following is true?

    I G must be abelian.

    II G has a subgroup of order 17.

    III G must have at least one proper non-trivial normal subgroup.

(A) I

(B) II

(C) I, II

(D) II, III

(E) I, II and III

Answer is (D). The dihedral group of a "17-gon" is a non-abelian group of order 34, so I is false. The 17-Sylow subgroup of G is of order 17. It is also of index 2, making it a proper non-trivial normal subgroup of G. So both II, and III are true.

14). The following is a sketch of the graph of $f(x) = x^3 + ax^2 + bx + c$.

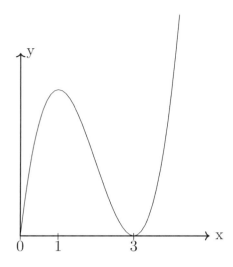

What is the value of $a + b + c$?

(A) 0

(B) 3

(C) 6

(D) 9

(E) 12

Answer is (B). We are given that $x = 0, 3$ are roots, and $x = 1, 3$ are roots of the derivative of f. The quickest way to solve this is to notice that if $x = 3$ is a root of both $f(x)$ and $f'(x)$ then $(x - 3)^2$ divide $f(x)$ (a root with multiplicity 2). Since $x = 0$ is a root as well $x(x - 3)^2$ divides $f(x)$. Since both are monic polynomials, $f(x) = x(x - 3)^2 = x^3 - 6x^2 + 9x$. So $a + b + c = -6 + 9 + 0 = 3$.

15). Which of the following most closely represents the graph of a solution to the differential equation $\frac{dy}{dx} = xy$?

(A)

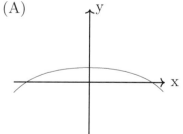

(B)

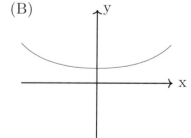

(C)

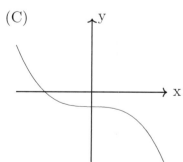

(D) 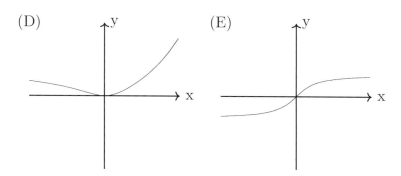 (E)

Answer is (B). The equation implies that the function $y(x)$ is increasing in the first and third quadrants, and decreasing in the second and forth.

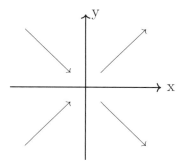

This helps us to eliminate (A) and (C). In addition, from the equation we can conclude that when x tends to $\pm\infty$ and as long as $y \neq 0$, the function exhibits extreme slope (either positive or negative) and cannot have any asymptotic behavior. The latter eliminates (D) and (E). For reference, a solution is $y = Ce^{x^2/2}$.

16). What is the minimal distance between the plane $x + 2y + 3z = 28$ and the sphere $x^2 + y^2 + z^2 = 2$?

(A) $\sqrt{14}$

(B) $2\sqrt{14}$

(C) $2\sqrt{14} - 1$

(D) $2\sqrt{14} - 2$

(E) $2\sqrt{14} - \sqrt{2}$

Answer is (E). The key consideration here is that due to the symmetry of the sphere we need to compute the distance of the plane from the origin and then subtract the radius $\sqrt{2}$ from the result (if we get a negative value, we conclude that they intersect). The bravest would choose (E) and move one on since it is the only possibility with $-\sqrt{2}$ in it. It is possible to solve it using Lagrange multipliers by minimizing the square of the distance $x^2 + y^2 + z^2$ subjected to $x + 2y + 3z - 28 = 0$. However, the clever thing to do here is to realize that we will get the minimum distance from the origin if the line connecting the origin to a point on the plane is perpendicular to the plane, i.e., the point must be on $(t, 2t, 3t)$. Plug in the plane equation to get: $t + 4t + 9t = 28 \Rightarrow (x, y, z) = 2(1, 2, 3)$. So the distance from the origin is $2 \cdot \sqrt{1 + 4 + 9} = 2\sqrt{14}$. Therefore the answer is $2\sqrt{14} - \sqrt{2}$.

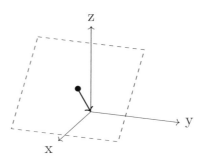

17). Let T be an isosceles triangle with base and height of length 2. Let $ABCD$ be a (convex) rectangle where two adjacent vertices, A and B, are located on T's base, C is located on one of T's legs and D in located on the other leg. What is the maximum possible area of $ABCD$?

(A) 0

(B) $\frac{1}{2}$

(C) 1

(D) 2

(E) None of the above.

Answer is (C). Let us use the following model:

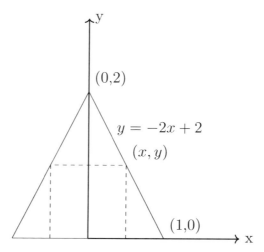

We wish to maximize $2xy = 2 \cdot x(-2x + 2)$ which is a downwards parabola with roots at $x = 0, 1$ so its maximal value is obtained at $x = 0.5$. Plug in to get a maximum area of 1.

18). Let $F(x) : (0, \infty) \to \mathbb{R}$ defined as follows:

$$F(x) = \int_x^{x^2} \frac{\log(t)}{t} dt.$$

Which of the following is true?

   I $\lim_{x \to \infty} F(x)$ exists and is bounded.

   II $\lim_{x \to \infty} \frac{dF}{dx}$ exists and is bounded.

   III $\lim_{x \to 0^+} \frac{dF}{dx}$ exists and is bounded.

(A) I

(B) II

(C) III

(D) I, II

(E) I, II and III

Answer is (B). By the Leibniz integral rule $F'(x) = 2x \frac{log(x^2)}{x^2} - \frac{\log(x)}{x} = 3 \frac{\log(x)}{x}$. So the derivative has a limit 0 at $x \to \infty$ but it is unbounded around 0 (use "one sided L'Hopital"). Thus II is true, but not III. Since the derivative is 0 at ∞, it is tempting to conclude that $F(x)$ is bounded, but we are familiar with counter examples for that misconception, such as $g(x) = \log(x)$. We can compute the integral explicitly:

$$I(y) = \int \frac{\log(t)}{t} dt = \int \frac{1}{t} \log(t) dt = \log(y)\log(y) - \int \log(t) \frac{1}{t} dt = \log^2(y) - I(y)$$

$$\Rightarrow I(y) = \frac{\log^2(y)}{2} \Rightarrow F(x) = \frac{\log^2(x^2)}{2} - \frac{\log^2(x)}{2} = \frac{3}{2} \log^2(x)$$

Now it is clear that $F(x)$ is unbounded.

19). Let $g(x) = |x - 1| + |x - 2|$.

$$\int_0^3 g(x) dx = ?$$

(A) 0

(B) 1

(C) 2

(D) 2.5

(E) 5

Answer is (E). This is an elementary question that can be time consuming. However if we sketch $|x - 1|$ and $|x - 2|$ we can see that they have the identical area under their graphs in $[0, 3]$ which composed of two triangles, one of area 0.5 and the other of area 2. So together g has area $2.5 \cdot 2 = 5$ under its graph.

20). Which of the following is FALSE?

    I A continuous map is an open map.

    II An open map is a continuous map.

    III A closed map is an open map.

(A) I

(B) II

(C) III

(D) I, II

(E) I, II and III

Answer is (E).
Counterexample for I: the point map $f : \mathbb{R} \to \mathbb{R}, x \mapsto 0$ with the standard topology is continuous but the image of every set is $\{0\}$ so it is not open.
Counterexample for II: The map $f : \mathbb{R} \to \hat{\mathbb{R}}, x \mapsto x$ where the domain has the standard topology and the image has the discrete topology has the required properties. Every map to a space with the discrete topology is open. However the preimage of the open set $\{0\} \subset \hat{\mathbb{R}}$ is $\{0\} \subset \mathbb{R}$ which is not open so the map is not continuous.
Counterexample for III: $f : [-1, 1] \to \mathbb{R}, x \mapsto x^2$. This is closed map since it is continuous on a compact set: every closed subset C of $[-1, 1]$ is closed and bounded in $\mathbb{R}$ so it is compact, therefore $f(C)$ is compact in $\mathbb{R}$, thus closed. However, this is not an open maps since $f(-1, 1) = [0, 1)$ which is not open in $\mathbb{R}$. 3 Note that the point map serves as a counter example for III as well, but the counterexample we use is more "pedagogical".

21). Let (a_n) be a sequence or real numbers. Which of the following is true?

(A) If $\lim_{n \to \infty}(a_n) = 1$ then it has a subsequence which converges to 0.

(B) If $\lim_{n \to \infty}(a_n) = 0$ then $\sum_{n=0}^{\infty} a_n$ converges.

(C) If $\lim_{n \to \infty}(a_n) = 0$ then there exists a subsequence (a_{n_k}) of (a_n) such that $\sum_{k=0}^{\infty} a_{n_k}$ is convergent.

(D) $\{a_n\}$ is a closed set of $\mathbb{R}$.

(E) None of the above is a true statement.

Answer is (C). (A) is false by taking $a_n = 1$. (B) is false by taking $a_n = \frac{1}{n}$. (D) is false by taking (a_n) to be the rational numbers (under some kind of enumeration). If it were closed then $\mathbb{R} \setminus \mathbb{Q}$ were open, but it does not contain any interval. Finally, (C) is true and we can construct it using the following method. Define $\epsilon_k = \frac{1}{2^k}$. Since $a_n \to 0$, then a_n has a infinitely many indices p for which $a_p \in [-\frac{1}{2^k}, \frac{1}{2^k}]$. Pick n_k out of those indices such that $n_k > n_{k-1}$. So we get a series for which $\sum_{k=0}^{\infty} |a_{n_k}| \leq \sum_{k=0}^{\infty} \frac{1}{2^k}$. So by the comparison test, the series converges absolutely and thus convergent.

22). A k-row is composed of a row of k (ordered) numbers $(x_1, x_2, ..., x_k)$ where the number in each slot is an integer from 1 to 10, and any number appears only once (i.e. there are no 11-row). A well-ordered k-row is a k-row in which for $i > j, x_i > x_j$. (i.e. (1,2,4) is a well ordered 3-row, but (1,4,2) is not). Consider a process in which a 7-row is chosen randomly. What is the probability to choose a well ordered 7-row?

(A) $\frac{1}{7!}$

(B) $\frac{1}{7!3!}$

(C) $\frac{1}{\binom{10}{7}}$

(D) $\frac{1}{\binom{10}{3}}$

(E) None of the above.

Answer is (A). For any well-ordered 7-row there are (7!-1) ways to permute the numbers to get a unique non-well-ordered 7-row. Another way to get the answer is to compute the cardinality of each set. The set of all 7-rows is of cardinality $10 \cdot 9 \cdot ... \cdot 4 = \frac{10!}{3!}$, whereas the set of well-ordered 7-rows can be counted as starting from the single well ordered 10-row and removing 3 numbers, so we get $\binom{10}{3} = \frac{10!}{7!3!}$. The answer follows.

23). Which of the following is true for a function $f(x) : (0, \infty) \to \mathbb{R} \setminus \{0\}$?

I $\lim_{x\to\infty} f = 0$ implies that $\frac{1}{f}$ is not bounded.

II $\lim_{x\to\infty} \frac{1}{f} = \infty$ implies that f is bounded.

III $\frac{1}{f}$ is unbounded in every interval containing $x = 3$ implies that $\lim_{x\to 3} f = 0$

(A) I

(B) II

(C) III

(D) I, III

(E) I, II and III

Answer is (A). I is true since if $\frac{1}{f}$ were bounded by M then $f(x) \geq \frac{1}{M}$ for any x and then the limit cannot be 0 at infinity. II is false by taking $f(x) = \frac{1}{x}$. III is false by taking f to be:

$$f(x) = \begin{cases} x & x \in \mathbb{Q} \\ x - 3 & \text{Otherwise} \end{cases}$$

Since 3 is rational, f does not obtain 0, however for any sequence of irrational numbers a_n tending to 3, $f(a_n) \to 0$. On the other hand, a rational sequence b_n tending to 3 will admit $f(b_n) \to 3$. So the limit of f do not exists at 3, although $\frac{1}{f}$ is unbounded around 3 by considering the same irrational sequence a_n.

24). Let $f : X \to X$ be a function for a set X to X. Let $A \subseteq X$. Denote $B = f(f^{-1}(A))$, $C = f^{-1}(f(A))$. Which of the following is true?

(A) $A \subseteq B \subseteq C$

(B) $A \subseteq C \subseteq B$

(C) $B \subseteq A \subseteq C$

(D) $C \subseteq A \subseteq B$

(E) $B \subseteq C \subseteq A$

Answer is (C). Since the question is general, we better take a simple example. Let $X = \{0, 1, 2\}, A = \{0, 1\}$ and let f be the point map to 0. So $B = f(f^{-1}(A)) = f(X) = \{0\}, C = f^{-1}(f(A)) = f^{-1}(0) = X$. So the answer follows.

25). What is the volume of the solid formed by revolving about the x-axis the region in the first quadrant of the xy-plane bounded by the coordinate axes and the the graph of $y = \frac{1}{x+1}$?

(A) π

(B) π^2

(C) 2π

(D) $\frac{\pi}{2}$

(E) ∞

Answer is (A). We need to preform the following calculation: $\pi \int_0^\infty y(x)^2 \, dx = \pi \int_0^\infty (x+1)^{-2} \, dx = -\pi(x+1)^{-1}|_0^\infty = \pi$.

26). Let A, B, C, D be 4×4 real matrices. Which of the following cannot be a true statement?

(A) The null space of BC is properly contained in the null space of ABC

(B) The null space of BC is properly contained in the null space of C

(C) The column space of AB is properly contained in the column space of CD

(D) The column space of BCD is properly contained in the column space of BC

(E) All of the above may be true.

Answer is (B). Setting A equals to the zero matrix, $D = C = B = I$, gives counterexamples for the statements (A) and (C). For (A), note that the null space of BC is trivial and the null space of ABC is $\mathbb{R}^4$. For (C), note that the column space of AB is trivial and the column space of CD is $\mathbb{R}^4$. For (D) let D be the zero matrix and the rest be I. The reason (B) must be false is because we have the reversed inclusion. If $Cv = 0$ then $BCv = B0 = 0$ so the null space of BC contains the null space of C.

27). Denote $\mathbb{R}[x]$ as the set of polynomials with real coefficients. Which of the following is a vector space?

   I $\{A \in \mathbb{C}^{n \times n} | A = -A^T\}$

   II $\{p(x) \in \mathbb{R}[x] | p(x) = 0, \text{ or } \deg(p(x)) \geq 2\}$.

   III $\{x \in \mathbb{C}^n | \exists \lambda \in \mathbb{C} \, s.t. \, Ax = \lambda x\}$, where $A \in \mathbb{C}^{n \times n}$ is fixed.

(A) I

(B) II

(C) III

(D) I, III

(E) I, II and III

Answer is (A).
For I: The set is the anti-symmetric matrices. This is a vector space as seen in the Linear Algebra module.
For II: The set is not closed under addition. $x^2 + (-x^2 + 1) = 1$ where 1 is not a polynomial of degree 2 or more nor the zero polynomial.
For III: This set is in fact the **union** (NOT sum) of the eigenspaces. It is not closed under addition in general. See the following example.

$$A = \begin{bmatrix} 1 & 0 \\ 0 & 2 \end{bmatrix},$$

$$A \begin{bmatrix} 1 \\ 0 \end{bmatrix} = \begin{bmatrix} 1 \\ 0 \end{bmatrix}$$

$$A \begin{bmatrix} 0 \\ 1 \end{bmatrix} = 2 \begin{bmatrix} 0 \\ 1 \end{bmatrix}$$

$$A \begin{bmatrix} 1 \\ 1 \end{bmatrix} = \begin{bmatrix} 1 \\ 2 \end{bmatrix} \neq \lambda \begin{bmatrix} 1 \\ 1 \end{bmatrix}$$

28). Let V be a vector space and $u, v, w \in V$ be vectors. Which of the following is true?

(A) If $\{u, v\}$ is a linearly dependent set and $\{u, w\}$ is a linearly dependent set, then $\{v, w\}$ is a linearly dependent set.

(B) If $\{u, v, w\}$ is a linearly dependent set then w is a linear combination of u and v.

(C) If $V = \text{Span}(\{u, v\})$ then V is of dimension 2.

(D) If V is of dimension 4, then exist a forth vector $x \in V$, such that $V = \text{Span}(\{u, v, w, x\})$

(E) None of the above.

Answer is (E). All the statements are false.
For (A), consider $V = \mathbb{R}^2, u = (0, 0), v = (1, 0), w = (0, 1)$.
For (B), consider again $V = \mathbb{R}^2, u = (0, 0), v = (1, 0), w = (0, 1)$.
For (C), consider $V = \mathbb{R}, u = 0, v = 1$.
For (D), consider $V = \mathbb{R}^4, u = v = w = (0, 0, 0, 0)$.

29). Let n be an integer bigger than 2 and let $f(x, y) = 3x^2 - 6xy + y^n$. Which of the following is true?

(A) f has no critical points for any value of n.

(B) Exists n such that f has exactly one critical point.

(C) Exists n such that f has a local minimum at the origin.

(D) Exists n such that f has a local minimum at point different than the origin.

(E) None of the above.

Answer is (D). We shall solve with the regular methods. First, $f_x = 6x - 6y \Rightarrow x = y$ is a solution to $f_x = 0$. Similarly, $f_y = -6x + ny^{n-1} \Rightarrow 6y = ny^{n-1} \Rightarrow y = 0$ and $y = \sqrt[n-2]{\frac{6}{n}}$ are solutions to $f_y = 0$ ($y = -\sqrt[n-2]{\frac{6}{n}}$ might be another solution if n is even). $f_{xx} = 6, f_{xy} = -6, f_{yy} = n(n-1)y^{n-2} \Rightarrow H = 6n(n-1)y^{n-2} - 36$. So at (0,0) we get $H < 0$ regardless of n, thus a saddle point. For the point $(\sqrt[n-2]{\frac{6}{n}}, \sqrt[n-2]{\frac{6}{n}})$ we get $H = 36(n - 2) > 0$. Since $f_{xx} = 6 > 0$ we get a minimum point at that critical point. So (D) is the right answer. Note: Sometimes a local extremum is called a "relative extremum".

30). Let P_1 be the plane $x + 2y - 4z = 0$. Let P_2 be the plane $10x + by + cz = 0$. Given that P_1 and P_2 are perpendicular to each other, and that their intersection is contained in the yz-plane. What is the value of $b + c$?

(A) 0

(B) 1

(C) -1

(D) 2

(E) None of the above.

Answer is (B). The line in the yz-plane has direction $(0, f, g)$ that is perpendicular to the normal of P_1, (1,2,-4). So we infer that the line is oriented in the direction of (0,2,1). Now the normal of P_2, (10,b,c), is perpendicular to both (1,2,-4) and (0,2,1). By comparing the dot product to zero we conclude $10 + 2b - 4c = 0$ and $2b + c = 0$. Solving for b, c we get $b = -1, c = 2$ and the answer follows.

31). Let $f(z) = f(x + iy) = u(x, y) + iv(x, y)$ be an entire complex function (analytic on the entire complex plane) and let $f'(z)$ denote its derivative. Given $f(0) = 0, f'(z) = 2z$. What is the value of $u(1, 2)$?

(A) 0

(B) 2

(C) -2

(D) 3

(E) -3

Answer is (E). By integrating we conclude that $f(z) = z^2 + c$. By Plugging in the the condition $f(0) = 0$ we get $f(z) = z^2 = x^2 - y^2 + i(2xy)$. So $u(x, y) = x^2 - y^2$ and the answer follows. Note that the antiderivative in complex analysis is unique up to a constant. An alternative solution is to use the property $f' = u_x + iv_x = v_y - iu_y = 2x + i2y$. So $u_x = 2x \Rightarrow u = x^2 + g(y) \Rightarrow u_y = g'(y) = -2y \Rightarrow u = x^2 - y^2 + c$.

32). Let G be the set of roots of $f(x) = x^7 - 1$ in the complex plane $\mathbb{C}$. Which of the following is FALSE?

(A) G is a multiplicative group with the multiplication operation from $\mathbb{C}$.

(B) G is an abelian group.

(C) G has a proper normal subgroup.

(D) $\prod_{g \in G} g = 1$.

(E) All the statements above are true.

Answer is (C). The roots from an multiplicative group of order 7, $G = \{e^{k \cdot 2\pi i/7} : k = 0, 1, 2, ..., 6\}$. Since the multiplication in $\mathbb{C}$ is commutative, the group is abelian. However, since its order is a prime number 7, it does not have a proper subgroups since for each subgroup $H < G$, the order $|H|$ divides 7. (D) is true by Vieta's formula: $x^7 - 1 = (x - g_0)(x - g_1) \cdot ... \cdot (x - g_6) \Rightarrow -1 = (-1)^7 \prod_{g \in G} g$.

Alternative explanation for (D): No elements of this group are self-inverse. If there were such a self-inverse element g, then the subgroup $\langle g \rangle$ would be of index 2, but 2 does not divide 7. Hence, we can group the elements of $G = \mathbb{Z}/7\mathbb{Z}$ into {1 identity element} and {3 pairs of mutually inverse elements}. The product then looks like $1 \cdot 1 \cdot 1 \cdot 1 = 1$ after multiplying each element with its (unique and different) inverse.

33). The following figure contains sketches of the graphs of $f(x) = x$, $g(x) = \frac{1}{x}$, $h(x) = \frac{x^2}{2}$:

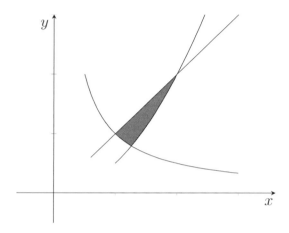

If A is the area of the gray region, what is the value of $3 - 6A$?

(A) $\log(2)$

(B) $\log(4)$

(C) $-\log(2)$

(D) $-\log(4)$

(E) None of the above.

Answer is (B). The intersection points are $(1, 1), (2^{1/3}, 2^{-1/3}), (2, 2)$. The two lower curves are for $\frac{1}{x}$ and $\frac{x^2}{2}$. The upper line is x. So we get:

$$A = \int_1^2 x\,dx - \int_1^{2^{1/3}} \frac{1}{x}\,dx - \int_{2^{1/3}}^2 \frac{x^2}{2}\,dx = \frac{3}{2} - \log(2^{1/3}) - 1 = \frac{3 - 2\log(2)}{6}$$
$$\Rightarrow 3 - 6A = 2\log(2) = \log(4)$$

34).

$$\lim_{N \to \infty} \sum_{k=1}^{N} \frac{\sin\left(\frac{k}{N}\pi\right)}{N} = ?$$

(A) 1

(B) 2

(C) $\frac{1}{\pi}$

(D) $\frac{2}{\pi}$

(E) $\frac{1}{2\pi}$

Answer is (D). This is a Riemann sum: we divide the interval $[0, 1]$ to N segments, $[0, 1/N], (1/N, 2/N], ..., ((N-1)/N, 1]$. For each segment we pick the rightmost point and sum up the rectangles area $\frac{1}{N} \sin(\frac{k}{N}\pi)$. If the following integral converges, the sum converges to $\int_0^1 \sin(x\pi) \, dx = \frac{1}{\pi} (-\cos(x\pi))\big|_0^1 = \frac{2}{\pi}$.

35). Observe the following function:

$$f(x) = \begin{cases} 0 & x \in \mathbb{Q} \\ 1 & \text{Otherwise} \end{cases}$$

What is the cardinality of the set of points of discontinuity of $f(x)$?

(A) 0

(B) 1

(C) countably infinite.

(D) uncountably infinite.

(E) None of the above.

Answer is (D). This is the famous Dirichlet function. It is discontinuous for any $x \in \mathbb{R}$ and therefore the answer is (D).

36). Let A be a matrix with characteristic polynomial $c_A(x) = x^3 - 6x^2 + 11x - 6$. What is the trace of A^{-1}?

(A) $\frac{1}{6}$

(B) $\frac{5}{6}$

(C) $\frac{11}{6}$

(D) 6

(E) None of the above

Answer is (C). The roots of $c_A(x)$ are the eigenvalues of A. By guessing the rational roots and dividing polynomials (refer to the basic module) one can find all the roots $x = 1, 2, 3$. Therefore A^{-1} must have $1, \frac{1}{2}, \frac{1}{3}$ as eigenvalues. Since the trace of a matrix is the sum of its eigenvalues, the answer is $1 + \frac{1}{2} + \frac{1}{3} = \frac{11}{6}$.

37). Let $A = \begin{pmatrix} 0 & 1 \\ 1 & 0 \end{pmatrix}$, and define $T : \mathbb{R}^{2\times 2} \to \mathbb{R}^{2\times 2}$ to be the linear transformation $T(X) = XA - AX$. Let U be the subspace of $\mathbb{R}^{2\times 2}$ spanned by $\left\{ \begin{pmatrix} 1 & 0 \\ 0 & 0 \end{pmatrix}, \begin{pmatrix} 0 & 1 \\ 0 & 0 \end{pmatrix} \right\}$. Let L be the restriction of T to U. What is the value of $\dim \text{Im} T - \dim \text{Im} L$?

(A) 0

(B) 1

(C) 2

(D) 3

(E) 4

Answer is (A). Let us inspect the the general term of $\mathrm{Im}T$:

$$T\begin{pmatrix} a & b \\ c & d \end{pmatrix} = \begin{pmatrix} a & b \\ c & d \end{pmatrix}\begin{pmatrix} 0 & 1 \\ 1 & 0 \end{pmatrix} - \begin{pmatrix} 0 & 1 \\ 1 & 0 \end{pmatrix}\begin{pmatrix} a & b \\ c & d \end{pmatrix}$$
$$= \begin{pmatrix} b & a \\ d & c \end{pmatrix} - \begin{pmatrix} c & d \\ a & b \end{pmatrix} = \begin{pmatrix} b-c & a-d \\ d-a & c-b \end{pmatrix}$$

We conclude that $\mathrm{Im}T$ is a vector space of dimension 2, since it is spanned by $\left(\begin{smallmatrix} 1 & 0 \\ 0 & -1 \end{smallmatrix}\right), \left(\begin{smallmatrix} 0 & 1 \\ -1 & 0 \end{smallmatrix}\right)$. In addition, by restricting T to U, i.e. setting $c = d = 0$, we get the same image. So the answer is $2 - 2 = 0$.

38). Let $\mathbf{F}$ be a vector field in $\mathbb{R}^3$, $\mathbf{F} = (x^2, -xy, (y+1)z)$. What is the work done by $\mathbf{F}$ on a particle that moves a along the path $(-t, t, \log(t+1))$ between time $t = 0$ and time $t = 1$?

(A) $\log(2)$

(B) $\log(4)$

(C) $-\log(2)$

(D) $-\log(4)$

(E) $\log(4) - 1$

Answer is (E). Do not be intimidated from the word "work". It simply means a line integral of the second type. Moreover, being in $\mathbb{R}^3$ is very similar to the $\mathbb{R}^2$ case. The work is:

$$\int_{path} x^2\,dx - xy\,dy + (y-1)z\,dz =$$

$$\int_0^1 t^2 \cdot -1 - (-t^2) \cdot 1 + (t+1)\log(t+1) \cdot 1/(t+1)\,dt =$$

$$\int_0^1 \log(t+1)dt = (t+1)\log(t+1) - t \,\vert_0^1 = 2\log(2) - 1 = \log(4) - 1$$

39). What is the global maximum of the function $f(x) = x^2 + y$ in the region $x^2 + y^2 \leq 1$?

(A) 0

(B) 0.5

(C) 1

(D) 1.25

(E) 1.5

Answer is (D). First notice that $f_y = 1 \neq 0$ and therefore the function does not admit local extrema. So left to check on the boundary using Lagrange multipliers:

$$F(x, y, \lambda) = x^2 + y - \lambda(x^2 + y^2 - 1)$$
$$F_x = 0 = 2x - \lambda 2x \Rightarrow 2x(1 - \lambda) = 0$$
$$F_y = 0 = 1 - \lambda 2y \Rightarrow 1 = 2\lambda y$$
$$\text{Case I: } x = 0 \Rightarrow y = \pm 1 \Rightarrow f = 1 \text{ at most.}$$
$$\text{Case II: } \lambda = 1 \Rightarrow y = 0.5 \Rightarrow$$
$$0 = F_\lambda \Rightarrow x^2 + 0.5^2 = 1 \Rightarrow x^2 = 0.75 \Rightarrow$$
$$f = 0.75 + 0.5 = 1.25$$

40). Let $\lfloor x \rfloor$ denote the greatest integer not exceeding x.

$$\sum_{k=0}^{\infty} (-1)^{\lfloor \frac{k}{2} \rfloor} \frac{\pi^k}{k!} = ?$$

(A) 0

(B) 1

(C) -1

(D) π.

(E) $-\pi$.

Answer is (C). The series is

$$1 + \pi - \frac{\pi^2}{2!} - \frac{\pi^3}{3!} + \frac{\pi^4}{4!} + \frac{\pi^5}{5!} - \frac{\pi^6}{6!} - \frac{\pi^7}{7!} + ...,$$

which is exactly the sum of the seires of $\sin(x)$ (odd powers) and $\cos(x)$ (even powers) evaluated at π. So the value is $\sin(\pi) + \cos(\pi) = 0 - 1 = -1$.

41). Let $A = \{0\} \cup \{1\} \cup (2, 4)$, $B = \{0\} \cup \{3\}$, be subsets of $\mathbb{R}$ with the standard topology, denoted as $(\mathbb{R}, T)$. Denote (A, T_A) the subset A with the subset topology. Which of the following is true? (Recall that y an isolated point of Y in (X, T_X) if there exists a neighborhood U of y in (X, T_X) such that $U \cap Y = \{y\}$.)

I A, as a subset of $(\mathbb{R}, T)$, has 2 isolated points.

II B, as a subset of (A, T_A), has 2 isolated points.

III B, as a subset of (A, T_A), has 1 interior point.

(A) I

(B) II

(C) III

(D) I, II

(E) I, II and III

Answer is (E). $0, 1$ are isolated points of A in $\mathbb{R}$ by taking small intervals around each. B as a subset of (A, T_A) has 2 isolated points by observing the neighborhoods $\{0\}$ and $(2, 4)$. Moreover, 0 is an interior point of B since B contains the neighborhood $\{0\}$, however any neighborhood of 3 is not contained entirely in B, so 3 is not an interior point of B.

42). Let A be $\begin{pmatrix} 1 & k & 2(1-k) \\ 0 & k & 1-k \\ 1 & k & 1-k \end{pmatrix}$, where $k \in \mathbb{C}$. Let b be $\begin{pmatrix} 0 \\ 0 \\ 1 \end{pmatrix}$. For which value of k, the system $Ax = b$ has infinitely many solutions?

(A) 1

(B) 0

(C) -1

(D) 0 and 1

(E) None of the above.

Answer is (E). After row reducing we get the following echelon form:

$$\left[\begin{array}{ccc|c} 1 & 0 & 0 & 1 \\ 0 & k & 0 & 1 \\ 0 & 0 & 1-k & -1 \end{array}\right]$$

$Rank(A|b) = 3 \Rightarrow$ We have only a single solution or none.

43). Which of the following is true for a group G?

I If G is of order 4, then G is abelian.

II If G is of order 6, then G is abelian.

III If all the subgroups of G are normal, then G is abelian.

(A) I

(B) II, II

(C) I,II

(D) I,III

(E) I,II and III

Answer is (A). There are only 2 groups of order 4, the cyclic $\mathbb{Z}_4$ and the Klein 4 group $\mathbb{Z}_2 \times \mathbb{Z}_2$. The dihedral group of the triangle is a non-abelian group of order 6. The quaternion group:

$$Q = \langle -1, i, j, k \mid i^2 = j^2 = k^2 = -1, ij = k, jk = i, ki = j \rangle$$

(as seen in the course) is a group of order 8 in which all the subgroups are normal. However, it is not abelian. $ij = k, ji = -k$.

44). Denote $d = \gcd(330, 105), m = \mathrm{lcm}(330, 105)$. What is $\gcd(d, m)$?

(A) 5

(B) 3

(C) 15

(D) 330

(E) 462

Answer is (C). First, factor 330 and 105 into prime factors: $330 = 3 \cdot 110 = 3 \cdot 2 \cdot 55 = 3 \cdot 2 \cdot 5 \cdot 11$ and $105 = 5 \cdot 21 = 5 \cdot 3 \cdot 7$. By comparing the prime factors, $d = 5 \cdot 3 = 15$. Since we know that the gcd divides the lcm we conclude that $\gcd(d, m) = d = 15$. Note: It is possible to find d using the Euclidian algorithm. However it is more time consuming.

45). Which of the following is the best approximation of $(8008)^{\frac{2}{3}}$?

(A) 330

(B) 400

(C) 460

(D) 550

(E) 1000

Answer is (B).

$$(8008)^{\frac{2}{3}} = (8 \cdot (10^3 + 1))^{\frac{2}{3}} = 4 \cdot (10^3 + 1)^{\frac{2}{3}} \approx 4 \cdot (10^3)^{\frac{2}{3}} = 400$$

46). Consider the the following statement:"Let a_n and b_n be sequences of positive real numbers. If $\lim_{n \to \infty} \frac{a_n}{b_n} = 1$, then $\lim_{n \to \infty}(a_n - b_n) = 0$."

Consider the following attempt to prove this statement:

(1) $\lim_{n \to \infty} \frac{a_n}{b_n} = 1$, then for any $\epsilon > 0$ exists N such that for $n > N$:

$$\left| \frac{a_n}{b_n} - 1 \right| \leq \epsilon$$

(2) Thus for $n > N$:

$$|a_n - b_n| \leq \epsilon b_n$$

(3) Since the latter is true for an arbitrary ϵ, then when $N \to \infty$, $|a_n - b_n| \to 0$.

(4) Therefore, $\lim_{n \to \infty}(a_n - b_n) = 0$

Which of the following is True?

(A) The argument is valid

(B) The inference in step (1) is wrong.

(C) The inference in step (2) is wrong.

(D) The inference in step (3) is wrong.

(E) The inference in step (4) is wrong.

Answer is (D). The inference in step (1) is true since it is merely the definition of a limit of a sequence. Since $b_n > 0$ for each n, the inference in step (2) is true. In step (4) we apply a known result, that a sequence tends to zero if and only if its absolute value tends to zero. However, the inference in step (3) is wrong because ϵb_n is not a constant that can be used for all n. Rather, ϵb_n varies by definition for each $n > N$ and may get arbitrarily large as $n \to \infty$ because b_n may be unbounded. A counterexample for the statement is this: $a_n = n^2 + n, b_n = n^2$. $\frac{a_n}{b_n} = 1 + \frac{1}{n} \to 1, a_n - b_n = n \to \infty$.

47). Let $C = \{z : |z| = 1\}$ by the unit circle in the complex plane, oriented counter-clockwise.

$$\oint_C \frac{\sin(z) + \cos(z)}{z} \, dz = ?$$

(A) $2\pi i$

(B) $4\pi i$

(C) πi

(D) $\frac{1}{2}\pi i$

(E) None of the above.

Answer is (A). By observing the power series for both $\sin(z) = z - \frac{z^3}{3!} + ...$ and $\cos(z) = 1 - \frac{z^2}{2!} + ...$ one see that $z = 0$ is the only pole and $Res(0, \frac{\sin(z)}{z}) = 0, Res(0, \frac{\cos(z)}{z}) = 1$. So by the residue theorem $\oint_C \frac{\sin(z)+\cos(z)}{z} dz = 2\pi i(0+1) = 2\pi i$.

48). By definition, a function $f(x) : \mathbb{R} \to \mathbb{R}$ is uniformly continuous if for any $\epsilon > 0$, exists $\delta > 0$ such that $|x - y| < \delta$ implies $|f(x) - f(y)| < \epsilon$. Which of the following statements is the negation to the statement of the latter definition?

(A) $\exists \epsilon > 0$ for which there exists $\delta > 0$ such that $|x-y| < \delta$ implies $|f(x)-f(y)| \geq \epsilon$

(B) $\forall \epsilon > 0, \exists \delta > 0$ such that $|x - y| < \delta$ implies $|f(x) - f(y)| \geq \epsilon$

(C) $\exists \epsilon > 0$ for which $\forall \delta > 0$, $|x - y| < \delta$ implies $|f(x) - f(y)| \geq \epsilon$

(D) $\exists \epsilon > 0$ for which $\forall \delta > 0$ there exists $x, y \in \mathbb{R}$ with $|x-y| < \delta$ and $|f(x)-f(y)| \geq \epsilon$

(E) None of the above.

Answer is (D). This is a logic question. The definition is of the form $\forall \epsilon, \exists \delta(\epsilon), \forall |x - y| < \delta, |f(x) - f(y)| < \epsilon$. In the negation, $\forall$ and $\exists$ switch rules, and $<$ turns to $\geq$ in the conclusion. $\exists \epsilon, \forall \delta, \exists |x - y| < \delta, |f(x) - f(y)| \geq \epsilon$.

49). Observe the following power series:

$$\sum_{k=0}^{\infty} \frac{(kx)^k}{k!}$$

What is the radius of convergence?

(A) 0

(B) ∞

(C) e

(D) $\frac{1}{e}$

(E) None of the above.

Answer is (D). By Cauchy-Hadamard / Ratio-Test

$$\frac{1}{R} = \lim_{k\to\infty} \left| \frac{a_{k+1}}{a_k} \right| = \lim_{k\to\infty} \left| \frac{(k+1)^{k+1}k!}{k^k(k+1)!} \right| = \lim_{k\to\infty} \left| \frac{k+1}{k} \right|^k = e \Rightarrow R = e^{-1}$$

50). Let $\mathbb{N}$ be the positive integers and define a function $f : \mathbb{N} \times \mathbb{N} \to \{0, 1, ..., 9\} \times \{0, 1, ..., 4\}$ by

$$f(a, b) = (a \mod 10, (a + b) \mod 5).$$

What is the minimum number m, such that for any $A \subset \mathbb{N} \times \mathbb{N}$ with cardinality m there exists $(x, y) \in \mathbb{N} \times \mathbb{N}$ with $|f^{-1}(x, y) \cap A| > 4$?

(A) 36

(B) 51

(C) 145

(D) 151

(E) 201

Answer is (E). The set of all possible images of $(a, b) \in \mathbb{N} \times \mathbb{N}$ is $B = \{0, 1, ..., 9\} \times \{0, 1, ..., 4\}$, which is of cardinality $10 \cdot 5 = 50$. The map f is surjective, meaning it obtains all those images; for a given $(x, y) \in B$, $f(x, y - x + 10) = (x, y)$. By the pigeon-hole principle, for any A with $50 \cdot 4 + 1 = 201$ elements, there is at least one subset of 5 elements in A that are mapped to the same element in B under f.

51). Which of the following polynomials has a root in neither the field of 3 elements $\mathbb{F}_3$, nor the field of 5 elements $\mathbb{F}_5$?

(A) $x^2 + x + 1$

(B) $x + 1$

(C) $x^2 + 1$

(D) $x^2 + 2x + 1$

(E) $x^2 + x + 2$

Answer is (E). Since we are dealing with small finite fields, we simply need to plug in all the possibilities for a root. In $\mathbb{F}_3$: 1 is a root of (A), 2 is a root of (B), and 2 is a root of (D) In $\mathbb{F}_5$, 2 is a root of (C). As for (E), plug in $x = 0, 1, 2, 3, 4$ compute the result in $\mathbb{R}$ and then view the result modulo 3 and modulo 5. This makes the computation easier. If $f(x) = x^2 + x + 2$ then $f(0) = 2$, $f(1) = 4$, $f(2) = 8$, $f(3) = 14$, $f(4) = 22$. Therefore f has no roots in either of these fields.

52). Let C be an ellipse in the xy-plane passing through $(2, 0), (0, 3)$ and $(-2, 0)$, oriented counterclockwise.

$$\oint_C (2x \sin(y) - y)dx + (x^2 \cos(y) + x)dy = ?$$

(A) 0

(B) 6π

(C) 12π

(D) 18π

(E) 24π

Answer is (C). We use Green's theorem. The value of the integral of question is:

$$\oint_C (2x\sin(y) - y)dx + (x^2\cos(y) + x)dy = \iint_{\text{ellipse}} (x^2\cos(y) + x)_x - (2x\sin(y) - y)_y \, dA$$
$$= \iint_{\text{ellipse}} 2dA$$

Since the area of the ellipse is $\pi \cdot 3 \cdot 2$, the answer is $2 \cdot 6\pi = 12\pi$.

53). Let A be a matrix with characteristic polynomial $c_A(x) = x^3 + x^2 + 1$. What is the trace of the inverse of A^2?

(A) 0

(B) 1

(C) -1

(D) 2

(E) -2

Answer is (E). By Cayley-Hamilton, $0 = A^3 + A^2 + I \Rightarrow -I = A^2(A + I)$. Then the inverse of A^2 is $-A - I$. So $\text{tr}(-A - I) = -\text{tr}A - 3 = 1 - 3 = -2$, where $\text{tr}(A) = -1$ since 1 is the coefficient of x^2 in $C_A(x)$.

54). Let $A \in \mathbb{C}^{n \times n}$ be a square complex matrix. Which of the following DO NOT imply that A is invertible?

(A) $Ax = 0$ has a single solution, for $x \in \mathbb{C}^n$.

(B) The columns of A are linearly independent.

(C) The characteristic polynomial of A is $x^5 + x^3 + x + 15$.

(D) A is diagonalizable, and its trace is 3.

(E) None of the above.

Answer is (D). Both (A) and (B) imply that the rank of A is n. (C) implies that the determinant of A is $-15 \neq 0$. Counter example for (D):

$$\begin{bmatrix} 0 & 0 & 0 \\ 0 & 0 & 0 \\ 0 & 0 & 3 \end{bmatrix}$$

55). Which of the following imply that a measurable set $A \subset \mathbb{R}$ has a positive measure?

(A) A is uncountable.

(B) A is dense.

(C) A is compact.

(D) $\mathbb{R} \setminus A$ is compact

(E) None of the above.

Answer is (D). The cantor set is a counterexample for (A). It is uncountable yet of measure 0. For (B): $\mathbb{Q}$ are dense in $\mathbb{R}$ but have measure 0. For (C): a single point is closed and bounded therefore compact, but of measure 0. (D) is correct, if $\mathbb{R} \setminus A = A^c$ is compact it is bounded, thus exists N s.t. $A^c \subset [-N, N]$ so $[N+1, N+2] \subset A$ so the measure of A is at least 1.

56). Let a, b, c be the eigenvalues of:

$$A = \begin{bmatrix} i & 1 & 1 \\ 1 & i & 1 \\ 1 & 1 & i \end{bmatrix}$$

What is the value of $(a+1)(b+1)(c+1)$?

(A) 0

(B) 1

(C) i

(D) $1 + i$

(E) None of the above

Answer is (E). Observe the following:

$$A = \begin{bmatrix} i & 1 & 1 \\ 1 & i & 1 \\ 1 & 1 & i \end{bmatrix} = \begin{bmatrix} 1 & 1 & 1 \\ 1 & 1 & 1 \\ 1 & 1 & 1 \end{bmatrix} + \begin{bmatrix} i-1 & 0 & 0 \\ 0 & i-1 & 0 \\ 0 & 0 & i-1 \end{bmatrix}.$$

So $A = B + (i-1)I$. Since B is of rank 1, and its trace is 3, the eigenvalues of B are $\{0, 0, 3\}$. Thus the eigenvalues of A are $\{i-1, i-1, i+2\}$. So $(a+1)(b+1)(c+1) = (i)(i)(i+3) = -3 - i$, which agrees with none of the possibilities.

57). Let $f : \mathbb{R} \to \mathbb{R}$ be a continuous function. Which of the following is true?

    I If $K \subset \mathbb{R}$ is compact, then $f(K)$ is compact.

    II If $K \subset \mathbb{R}$ is compact, then $f^{-1}(K)$ is compact.

    III If $K \subset \mathbb{R}$ is connected, then $f(K)$ is connected.

(A) I

(B) I, II

(C) II, III

(D) I, III

(E) I, II and III

Answer is (D). I and III are well known theorems. A counter example for II would be the point map $f(x) = 0$. Then $\{0\}$ is compact but $f^{-1}(\{0\}) = \mathbb{R}$ is not.

58). Consider the sequence of real functions: $f_n(x) = \sin^n(x)$. Consider the following intervals: $I_1 = (-4, \frac{\pi}{2}], I_2 = (-\frac{\pi}{2}, \frac{\pi}{2}), I_3 = (-\frac{\pi}{2}, \frac{\pi}{2}], I_4 = (-\frac{3}{2}, \frac{3}{2}]$. We say that an interval I_j is "bigger" than I_k if $I_j \supset I_k$. Which of those intervals is the biggest interval in which f_n converges uniformly?

(A) I_1

(B) I_2

(C) I_3

(D) I_4

(E) None of the above.

Answer is (D). The limit function of f_n is:

$$f(x) = \begin{cases} 1 & x = \frac{\pi}{2}(2k+1), k \in \mathbb{Z} \\ 0 & \text{Otherwise} \end{cases}$$

Moreover, if $f_n(x) \to f(x)$ uniformly on (a, b), and $f_n(x)$ are all continuous on (a, b) then $f(x)$ must be continuous on (a, b) as well. This criterion eliminates (A). We can guess that f_n would behave "non-uniformly" at $x = \frac{\pi}{2}$ just as x^n behaves at $x = 1$. And indeed for $\epsilon = 0.1$, for any N, we can pick $x < \frac{\pi}{2}$ such that $\sin(x) = 2^{-\frac{1}{N}}$ so $f_N(x) = 0.5$. Therefore, $f_n(x)$ does not converge uniformly on $(-\frac{\pi}{2}, \frac{\pi}{2})$. In order for it to converge uniformly, the interval must be well inside $(-\frac{\pi}{2}, \frac{\pi}{2})$, namely of the form $[-\frac{\pi}{2} + \delta, \frac{\pi}{2} - \delta]$. That is because in such interval, $|f_n(x)|$ is bounded by $\sin(\frac{\pi}{2} - \delta)^n$, which is convergent to 0 since $\sin(\frac{\pi}{2} - \delta) < 1$. Thus for any $\epsilon > 0$ we have N s.t. for $n > N$ $|f_n(x)| < \sin(\frac{\pi}{2} - \delta)^N < \epsilon$ regardless of x. I_4 fits that description and it is the right answer.

59). Let M be a module over the ring $\mathbb{Z}_5$, which denotes the set $\{0, 1, 2, 3, 4\}$ with addition and multiplication modulo 5. Which of the following must be true?

(A) M is finite.

(B) M is a vector space.

(C) If M is a ring, the multiplication in M is commutative.

(D) M must be isomorphic to $\mathbb{Z}_5$.

(E) None of the above must be true.

Answer is (B). Since 5 is prime, $\mathbb{Z}_5$ is a field (commutative ring with inverse for every non zero element). Every module over a field is a vector space by definition. The polynomials with coefficients in $\mathbb{Z}_5$ form an example for a $\mathbb{Z}_5$-module which is not finite, and not isomorphic to $\mathbb{Z}_5$, eliminating (A) and (D). 2 by 2 Matrices with entries in $\mathbb{Z}_5$ are an example for a non-commutative ring which is also a module over $\mathbb{Z}_5$:

$$\begin{bmatrix} 1 & 0 \\ 0 & 0 \end{bmatrix} \begin{bmatrix} 0 & 1 \\ 1 & 0 \end{bmatrix} = \begin{bmatrix} 0 & 1 \\ 0 & 0 \end{bmatrix}$$

$$\begin{bmatrix} 0 & 1 \\ 1 & 0 \end{bmatrix} \begin{bmatrix} 1 & 0 \\ 0 & 0 \end{bmatrix} = \begin{bmatrix} 0 & 0 \\ 1 & 0 \end{bmatrix}$$

60). A **complete** graph of n vertices, denoted K_n, is an undirected graph with n vertices and a unique edge connecting each pair of vertices. An **Eulerian** cycle is a cycle passing through every edge exactly once. Let $E(n)$ denote the number of edges in K_n. For which n, K_n has an Eulerian cycle and $E(n)$ is divisible by 5?

(A) 4

(B) 5

(C) 6

(D) 10

(E) 13

Answer is (B). A necessary and sufficient condition for the existence of an Eulerian cycle is that for any vertex, the number of edges connected to it is even. Those who have not taken Graph Theory are expected to infer the "necessary" part as follows: Given such cycle, imagine walking on that cycle starting from an arbitrary vertex. Whenever we reach a vertex using one edge, we leave it using another. This is true for any of the vertices except for the starting point, however for it we say that we started by walking on one edge and ended the walk by using another. We conclude that the number of edges connected to every vertex is even.

For K_n, any vertex has $(n-1)$ edges connected to it, so we eliminate all the even n's: (A),(C),(D). The total number of edges is the number of pairs on n vertices: $E(n) = \binom{n}{2} = \frac{n(n-1)}{2}$. So $E(5) = 10$ and $E(13) = 13 \cdot 6 = 78$. Only (B) can be true, and indeed K_5 has an Eulerian cycle by "walking" on the pentagon first and then on the pentagram.

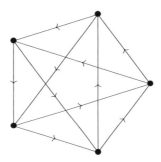

61). For any ring A, let $A[x]$ denote the polynomials with coefficients in A. Let A_k, where $k \in \mathbb{R}$, denote the set $\{k\} \times A$ (cartesian product). Let $\mathbb{Q}$ denote the rational numbers, and $\mathbb{N}$ - the natural numbers. Which of the following sets has a different cardinality than the others?

(A) $\bigcup_{k \in \mathbb{Q}} \mathbb{Q}_k$

(B) $\mathbb{Q}[x]$.

(C) $\mathbb{Z}[x]$

(D) $\mathbb{Q} \times \mathbb{Q}$ (cartesian product).

(E) The set of all functions from $\mathbb{N}$ to $\mathbb{N}$.

Answer is (E). (A),(B),(C) are all countable union of countable sets and therefore countable. (D) is a finite cartesian product of countable sets and thus countable as well. On the other hand, functions from $\mathbb{N}$ to $\mathbb{N}$ has cardinality of at least as the set of all function $\mathbb{N} \to \{0, 1\}$ which is the power set (the set of all subsets of $\mathbb{N}$) and thus is uncountable.

62).

$$i \tan(i) = ?$$

(A) $\frac{1-e^2}{1+e^2}$

(B) $\frac{1+e^2}{1-e^2}$

(C) $\frac{1-e}{1+e}$

(D) $\frac{1+e}{1-e}$

(E) $\frac{1+e^2}{1-e^2} i$

Answer is (A).

$$i \tan(i) = i\frac{\sin(i)}{\cos(i)} = i\frac{i \sinh(1)}{\cosh(1)} = -\frac{e - e^{-1}}{e + e^{-1}} = -\frac{e^2 - 1}{e^2 + 1} = \frac{1 - e^2}{1 + e^2}$$

63). How many abelian groups of order 540 do not have a subgroup isomorphic to the Klein four-group? (Only count the groups up to isomorphism.)

(A) 2

(B) 3

(C) 4

(D) 5

(E) 6

Answer is (B). By classification of abelian groups, the there are a total of 6 groups of order $540 = 2^2 \cdot 3^3 \cdot 5$.

$\mathbb{Z}_5$	$\oplus$	$\mathbb{Z}_2 \oplus \mathbb{Z}_2$	$\oplus$	$\mathbb{Z}_3 \oplus \mathbb{Z}_3 \oplus \mathbb{Z}_3$	
$\mathbb{Z}_5$	$\oplus$	$\mathbb{Z}_2 \oplus \mathbb{Z}_2$	$\oplus$	$\mathbb{Z}_{3^2} \oplus \mathbb{Z}_3$	
$\mathbb{Z}_5$	$\oplus$	$\mathbb{Z}_2 \oplus \mathbb{Z}_2$	$\oplus$	$\mathbb{Z}_{3^3}$	
$\mathbb{Z}_5$	$\oplus$	$\mathbb{Z}_{2^2}$	$\oplus$	$\mathbb{Z}_3 \oplus \mathbb{Z}_3 \oplus \mathbb{Z}_3$	
$\mathbb{Z}_5$	$\oplus$	$\mathbb{Z}_{2^2}$	$\oplus$	$\mathbb{Z}_{3^2} \oplus \mathbb{Z}_3$	
$\mathbb{Z}_5$	$\oplus$	$\mathbb{Z}_{2^2}$	$\oplus$	$\mathbb{Z}_{3^3}$	

The top 3 contains a copy of the Klein 4 group $\mathbb{Z}_2 \oplus \mathbb{Z}_2$, and the bottom 3 do not.

64). Let $A = \{a_n\}_{n \in \mathbb{N}}$ be a sequence of real numbers. Which of the following is true?

(A) A is closed.

(B) A is compact.

(C) If A is bounded, then A is compact.

(D) If $\lim_{n \to \infty} a_n$ does not exist, then A is closed.

(E) If $\lim_{n \to \infty} a_n = L$ And $L \in A$ then A is compact.

Answer is (E). Consider the following example.

$$a_n = \begin{cases} 1 - \frac{1}{n} & n \text{ is even} \\ \frac{1}{n} & \text{Otherwise} \end{cases}.$$

The limit does not exists, and A fails to contain all its limit points, so it is not closed, let alone compact, although it is bounded. This eliminates (A),(B),(C) and (D). (E) is true since the assumption suggest that A has only one limit point which it contains. In addition, a convergent sequence is bounded and thus A is compact.

65). Let $\mathbb{R}$ denote the real line with the standard topology. Let X denote the real line with the following topology: $U \subset X$ is open $\iff$ $X \setminus U$ is finite or $U = \emptyset$. Which of the following is true?

   I Exists a proper set $A \neq \emptyset$ which is compact in both X and $\mathbb{R}$.

   II Exists a proper set $A \neq \emptyset$ which is open in both X and $\mathbb{R}$.

   III Every subset of X is compact.

(A) I

(B) II

(C) III

(D) I, III

(E) I, II and III

Answer is (E). For a single point $\{a\}$ the set $\mathbb{R} \setminus \{a\}$ is open in both topologies, which proves II. The following proves III: for any given set A in X, a single open set U must cover all of A except at most finitely many points. So for a given open cover $\{U_\alpha\}$ of A arbitrarily pick one of the open sets, say U_1. We are left with N points in A not covered by U. So now pick N additional open sets, one for each of the N points. Together they form a finite sub-cover. So III is true, which implies I immediately.

66). Let $f(x) : \mathbb{R} \to \mathbb{R}$ be a continuous function. Given $f(-1) = 1, f(1) = 3$. Which of the following is the weakest requirement to ensure the existence of $c \in (-1, 1)$ with $\frac{df}{dx}(c) = 1$?

(A) f is differentiable on $[-1, 1]$ and $\frac{df}{dx}$ is bounded there.

(B) f is differentiable on $[-1, 1]$.

(C) f is differentiable on $(-1, 1)$ and $\frac{df}{dx}$ is continuous there.

(D) f is differentiable on $(-1, 1)$.

(E) f is differentiable on $\mathbb{R}$ and $\frac{df}{dx}$ is continuous.

Answer is (D). This is a direct application of the mean value theorem, which requires f to be continuous on [-1,1] and differentiable on (-1,1). The derivative may be unbounded and may not exists on $x = \pm 1$, (see example at the figure below). Also, the theorem does not require continuity of the derivative.

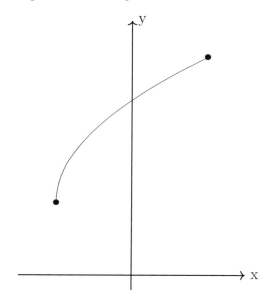

Made in the USA
Middletown, DE
10 August 2015